THE RED CELL MEMBRANE

Seminars in Hematology

THE RED CELL MEMBRANE

Edited by

Robert I. Weed, M.D.

Department of Internal Medicine,
University of Rochester School of Medicine and Dentistry,
Rochester, New York

Ernst R. Jaffé, M.D.

Department of Medicine,
Albert Einstein College of Medicine,
New York, New York

Peter A. Miescher, M.D.

Division of Hematology, Medical Service,
University of Geneva, Hôpital Cantonal,
Geneva, Switzerland

GRUNE & STRATTON
New York and London

The chapters of this book appeared originally in the July and October 1970 issues (Vol. VII, Nos. 3 and 4) of *Seminars in Hematology*, a quarterly journal published by Henry M. Stratton, Inc., and edited by Peter A. Miescher, M.D., and Ernst R. Jaffé, M.D. Robert I. Weed, M.D., was guest editor for these issues.

Grune & Stratton, Inc.
111 Fifth Avenue, New York, New York 10003

Library of Congress Catalog Card Number 70-170193
International Standard Book Number 0-8089-0736-0

Printed in the United States of America (PC-H)

Contents

Contributors

RICHARD A. COOPER, M.D., Thorndike Memorial Laboratory and Harvard Medical Unit, Boston City Hospital, Boston, Massachusetts.

HARRY S. JACOB, M.D., Chief, Section of Hematology and Professor of Medicine, University of Minnesota Medical School, Minneapolis, Minnesota.

WALLACE N. JENSEN, M.D., Professor and Chairman, Department of Medicine, The George Washington University, Washington, D.C.

P. L. LACELLE, M.D., Associate Professor of Medicine and of Radiation Biology and Biophysics, University of Rochester School of Medicine and Dentistry, Rochester, New York.

PETER K. LAUF, M.D., Assistant Professor of Physiology and Immunology, Duke University Medical Center, Durham, North Carolina.

LAWRENCE S. LESSIN, M.D., Division of Hematology, Department of Medicine, The George Washington University, Washington, D.C.

DR. A. H. MADDY, The Department of Zoology, University of Edinburgh, Edinburgh, Scotland.

N. SCOTT MCNUTT, M.D., Clinical and Research Fellow, Department of Pathology, Massachusetts General Hospital, Boston, Massachusetts; and Teaching Fellow, Harvard Medical School, Cambridge, Massachusetts.

DAVID G. NATHAN, M.D., Chief, Division of Hematology, Children's Hospital Medical Center; Assistant Professor of Pediatrics, Harvard Medical School, Boston, Massachusetts.

WENDELL F. ROSSE, M.D., Associate Professor of Medicine and Immunology, Duke University Medical Center, Durham, North Carolina.

STEPHEN B. SHOHET, M.D., Associate in Hematology, Children's Hospital Medical Center; Assistant Professor of Pediatrics, Harvard Medical School, Boston, Massachusetts; Fellow of the Medical Foundation, Inc.

MEHDI TAVASSOLI, M.D., Research Fellow, Department of Anatomy, The Johns Hopkins University School of Medicine, Baltimore, Maryland; Blood Research Laboratory, Tufts-New England Medical Center, Medford, Massachusetts.

ROBERT I. WEED, M.D., Professor of Medicine and of Radiation Biology and Biophysics; Head, Hematology Unit, Department of Internal Medicine, University of Rochester School of Medicine and Dentistry, Rochester, New York.

The affiliation listed here for each contributor is that given in the original journal publication of the symposium.

RONALD S. WEINSTEIN, M.D., Head, Mixter Laboratory for Electron Microscopy; and Clinical and Research Fellow, Massachusetts General Hospital, Boston, Massachusetts; and Teaching Fellow, Harvard Medical School, Cambridge, Massachusetts.

LEON WEISS, M.D., Professor of Anatomy, Department of Anatomy, The Johns Hopkins University School of Medicine, Baltimore, Maryland.

Preface

IN THE ORIGINAL INTRODUCTION to this symposium, which first appeared in the July and October 1970 issues of *Seminars in Hematology*, reference is made to the rapid expansion of knowledge and interest in disorders of the red cell membrane. This state of affairs has been further reflected by the continuing interest in the symposium on the part of hematologists in practice, those engaged in clinical investigation and those engaged in basic research studies. Requests for reprints have been so great in number that several of the participating authors have exhausted their supply, and no more copies of the journal issues themselves are available. This very positive response by many readers certainly constitutes a tribute to the contributing authors.

Grune & Stratton, together with the editors of the symposium, decided to issue *The Red Cell Membrane* in book form to meet this continuing demand. We hope that the collection of these contributions into a bound volume will provide a valuable reference source for students, investigators, clinical hematologists and those hematologists concerned with disorders of the red cell who divide their time between the laboratory and the clinic.

Robert I. Weed, M.D.,
Guest Editor

Disorders of the Red Cell Membrane: History and Perspectives

By Robert I. Weed

THE RAPID EXPANSION of knowledge and interest in disorders of the red cell membrane has stimulated publication of five previous reviews[24,58,61,38a,14] on the subject between 1965 and the present. In recognition of this rapidly developing field, this symposium not only includes discussions of structural or chemical abnormalities which have been identified in various disorders or groups of disorders but also includes comprehensive reviews of current methodology applicable to the study of erythrocytes in pathologic states. This issue of SEMINARS contains reviews of red cell membrane lipids and proteins by Drs. Cooper[5] and Maddy,[31] respectively, and of ultrastructure by Drs. Weinstein and McNutt.[64] Also included are discussions of antibody effects by Drs. Rosse and Lauf[48] and Heinz body disorders by Dr. Jacob.[22] The October issue of SEMINARS will contain discussions on altered cellular deformability by Dr. LaCelle,[29] the hazards of splenic passage by Drs. Weiss and Tavassoli,[62] cation transport and permeability by Drs. Nathan and Shohet[38] and a discussion of cellular fragmentation in hemoglobinopathies by Drs. Jensen and Lessin.[27]

Within the past two decades, biochemical advances have made it possible to identify molecular defects of human erythrocytes in a wide variety of anemic states. These advances can be grouped into three categories. Recognition of the electrophoretic abnormality of sickle cell hemoglobin[39] ushered in a highly productive period of research into normal and abnormal hemoglobin structure and function which continues to provide new insights into the pathology of abnormal hemoglobins. The recognition of glucose-6-phosphate dehydrogenase deficiency as the defect underlying primaquine-induced hemolysis[4] triggered a period of remarkable interest and intensive search by many laboratories for other enzymic deficiencies predisposing to

From the University of Rochester School of Medicine and Dentistry, Rochester, N.Y.

Supported in part by USPHS Research Grant HE-06241-09 and by the United States Atomic Energy Commission at the University of Rochester Atomic Energy Project, Rochester, N.Y. It has been assigned publication number UR-49-1253.

Robert I. Weed, M.D.: *Professor of Medicine and of Radiation Biology and Biophysics, Head, Hematology Unit, Department of Internal Medicine, University of Rochester School of Medicine and Dentistry, Rochester, N.Y.*

1

a shortened erythrocyte life span. Research on disorders of the red cell
membrane, the subject of this and the October issues of SEMINARS, really began
with descriptive morphologic observations. While it is difficult to identify the
first specific meaningful studies of a biochemical and structural defect in membranes, hematologists interested in hemolytic disease should properly be
credited with the earliest interest in disorders of the red cell membrane since
the event of hemolysis implies breakdown of the membrane, whether it occurs
in the test tube in vitro or within a phagocyte in vivo.

Just as an abnormality in the structure of hemoglobin and the metabolic
nature of the hemolytic defect in primaquine-induced hemolysis were suspected
before the specific observations were made, membrane alterations in vitro
were suggested by urea-induced fragmentation reported as early as 1864
by Kölliker. The pathophysiologic significance of membrane fragmentation
both in vitro and in vivo has been suggested by many workers during the
past century. If one defines fragmentation as loss of a portion of the cell
membrane which may or may not contain some hemoglobin, it is clear that
red cell fragmentation is a very important final common pathway for red
cell destruction in a variety of hemolytic states, occurring secondary to
both intrinsic red cell abnormalities and to alterations in the normal microcirculatory environment.

Table 1 summarizes some of the first or most important contributions in
the history of research and description of the phenomenon of red cell fragmentation. In every example cited in Table 1, hemolysis either in vitro or
in vivo, measured as shortened survival of some red cell label, has been
found to be the consequence of fragmentation. The work of Ganzoni et al.[15]
adds macroreticulocyte maturation to the variety of conditions accompanied
by fragmentation. Only within the last six to 8 years, however, has the
proper clinical significance been accorded erythrocyte fragmentation. In the
October issue Drs. Jensen and Lessin[27] will focus particular attention on the
importance of red cell fragmentation as a pathophysiologic mechanism in sickle
cell and hemoglobin CC disease.

Apart from the long history of work on fragmentation, application of
modern physiological and biochemical technics to the study of membrane
structure and function in pathologic red cells probably dates to the study
of monovalent cation fluxes in sickle cells by Tosteson in 1951.[55] From a
biochemical viewpoint, the demonstration that all or virtually all of the lipid
of mature erythrocytes is to be found in the membrane[57,10] has made it
clear that various disorders of the red cell characterized by qualitative, quantitative or dynamic alterations in lipids are, in fact, membrane disorders.
This subject is reviewed extensively by Dr. Cooper in this issue.[5]

Any history of biophysical studies of the red cell membrane in disease
states should certainly credit the earliest studies of mechanical and osmotic
fragility. Although these laboratory tests were not definitively interpretable
at the time that they were first found useful for evaluation and diagnosis of
clinical disease, their physiologic significance has now become much more
apparent. Stated in its simplest form, the osmotic fragility (O.F.) test provides a way of evaluating the surface area/volume (SA/V) relationship of

Table 1.—Red Blood Cell Fragmentation

Reference	Type
1864 Kölliker	Urea fragmentation
1865 Schultze[50]	Heat fragmentation
1884 Meltzer and Welch[33]	Glass bead fragmentation
1891 Ehrlich[11]	Schiztocytes in vivo
1917 Rous and Robertson[48]	Relation to normal destruction in vivo
1946 Brown[2]	Thermal injury
1947 Marmont and Bianchi[32]	Thalassemia
1948 Ham, Shen and Castle[49]	Thermal injury
1951 Bessis, Bricka and Dupuy[2]	Myelin forms from agglutinated red cells
1953 Policard and Bessis[40]	Partial phagocytosis of antibody injured cells
1953–54 Dacie et al.[6,7]	Hemolytic-uremic syndrome
1955 Gasser et al.[16]	
1954 Rose et al.[47]	Prosthetic heart valve hemolysis
1954–55 Ponder[41]	Urea, $(NH_4)_2SO_4$, heat (normal and thalassemic)
1963 Westerman, Pierce and Jensen[66]	Loss of membrane lipid with aging in vivo
1964 Koyama, Aoki and Deguchi[28]	Pitting of Heinz bodies within the spleen
1964 Rand and Burton[43,44]	Viscoelastic breakdown of membrane
1965 Fuhrmann[13]	Urea, ammonia
1965 Jensen and Bessis[26]	Sickle cell fragmentation
1965 Sears and Crosby[51]	Called attention to fragmentation in iron deficiency and hepatic necrosis
1965 Archer[1]	Role of monocytes in partial phagocytosis
1967 Lobuglio and Jandl[30]	
1966 Weed and Bowdler[59]	Loss of membrane from hereditary spherocytes
1966 Reed and Swisher[46]	
1966 Weed and Weiss[60]	Intrasplenic fragmentation of antibody injured cells
1968 Slater, Muir and Weed[52]	Intrasplenic fragmentation of β thalassemia and hemoglobin H cells
1968 Wennberg and Weiss[64]	
1969 Ginn, Hochstein and Trump[17]	"Internal" fragmentation produced by primaquine
1969 Haradin, Reed and Weed[19]	Membrane loss from ACD-stored RBC
1969 Ganzoni, Hillman and Finch[15]	Loss of hemoglobin during macro-reticulocyte maturation

normal and abnormal cells. Thus, increased fragility is generally interpretable as indicative of a decrease in SA/V or sphering. Conversely, decreased O.F. (increased resistance to lysis in hypotonic salt solutions) implies either an increase in surface area or a decrease in cation content or volume in an isotonic medium. Both hereditary spherocytosis and spherocytosis seen in immune hemolytic disease are associated with increased O.F., while the increase in membrane lipid content seen in a variety of disorders is associated with enhanced osmotic resistance, as reviewed by Cooper.[5] In spite, however, of broadened insights into the pathophysiologic alterations in red cells implied by altered O.F., it is still essential to recall the fundamental assumptions upon which the test is based. First, as a way of estimating red cell surface area, the O.F. test assumes a normal cell volume in plasma, reflecting primarily a normal total cation content which should not change in prelytic, hypotonic salt solutions. Inferences regarding decreased surface area from an increased O.F. must exclude any increase in cell volume. Interpretation of increased surface area based on decreased O.F. must also be based on normal volumes in plasma and must exclude any excessive prelytic loss of cation from partially swollen cells, as demonstrated by Ponder[41] in cells affected by certain detergents and surface active agents. Finally, interpretation of O.F. as an expression of SA/V assumes that the intrinsic properties of the membrane have not changed. In hereditary spherocytosis, however, the membrane itself becomes very rigid[29,59] and shows an anomalously low swelling prior to lysis and shrinkage after lysis.

Mechanical fragility, on the other hand, is related also to the shape factor, i.e., decreased SA/V (sphering) makes the cell less able to tolerate deforming stresses and, therefore, more mechanically fragile. In addition, however, mechanical fragility may reflect either a cell with rigid contents, e.g., a sickle cell, or cells which have lost their normal intrinsic membrane deformability, such as metabolically depleted cells.[56] Thus, rather than being a nonspecific indicator of abnormal red cells, the fact that mechanical fragility is abnormal in so many disorders strongly suggests that loss of cellular deformability for any reason, whether it be shape change or development of a rigid interior or membrane, is a final common pathway leading to premature cellular destruction.

Actually, in retrospect, recognition on stained blood films of a wide variety of morphologic abnormalities of the red cell has for a long time provided us with very important clues about the physical state of the red cell membrane and the contents contained therein since any significant deviation from the normal smooth contour of the biconcave disc suggests rigidity of the membrane or of the cellular contents.

The introduction by A. Teitel,[53] Jandl[23] and P. Teitel[54] of measurements of red cell filterability have provided a clearer analogue to the hazards of pasage for abnormal cells through the microcirculation and demonstrated that increases in cellular rigidity are associated with many hemolytic states. A very significant advance in the study of intrinsic cellular physical properties has been made possible by the work of Rand and Burton[43,44] who studied erythrocytes with the micropipette technic, originally employed by Mitchison and Swann,[34] to study sea urchin egg membranes. As Dr. LaCelle will discuss in the October issue of SEMINARS,[29] the micropipette not only permits cellular rigidity to be analyzed into membrane rigidity versus rigidity of contents but also permits evaluation of individual abnormal cells within a population whose mean may not be greatly different than normal. Survival of the red cell in vivo is dependent on preservation of both its normal biconcave shape and the intrinsic deformability of the membrane, as well as fluidity of cellular contents. The interrelation and dependence of cellular deformability upon those parameters will also be discussed by Dr. LaCelle in his review.[29]

Among the many questions dealt with in this symposium, attention is directed to the question of what are the specific determinants of red cell life span in vivo? Maintenance of those geometric and physical properties which are essential for normal cellular deformability, as will be discussed by Dr. LaCelle,[29] depends on the critical role of ATP in maintaining cellular shape[37] and preventing calcium-induced membrane rigidity,[56] thus providing one explanation why a variety of enzymic deficiencies in the Embden -Meyerhof pathway might readily predispose red cells to a shortened life span. In addition, when the physical state of intracellular hemoglobin is altered toward decreased solubility, as in the case of sickle hemoglobin, the result is to make the entire cell rigid, clearly predisposing to intravascular fragmentation and removal from the circulation as discussed in the review by Dr. Jensen.[27] The review by Dr. Jacob[22] focuses our attention on the unstable hemoglobin syndromes which are associated with the formation of Heinz bodies that may interact with the cell membrane to affect adversely its permeability properties as well as its deformability. Additional Heinz body disorders include precipitated alpha or beta chains in thalassemic syndromes[12,52,64] or glutathione-hemoglobin mixed disulfides[25] arising secondary to metabolic defects in the pentose phosphate shunt when challenged by oxidant drugs or chemicals.

In addition to changes in shape, such as the disc to sphere transformation or primary alterations in cellular or membrane rigidity, many hemolytic disorders have been shown to be associated with abnormalities of active and passive cation flux. Considerable controversy has centered around the pathophysiologic significance of these abnormalities in specific disorders such as hereditary spherocytosis. This subject will be reviewed in the October issue by Drs. Nathan and Shohet.[38] The observations made on pyruvate kinase deficient cells appear to provide us with an example of how increased rigidity of the cell may be related to loss of cation.

The manner in which the membrane is altered by important extrinsic determinants will also be reviewed in this symposium. Drs. Weiss and Tavassoli[63] will focus attention on the role of the ultrastructure of the bone marrow and spleen in the destruction of various kinds of rigid red cells, while Drs. Rosse and Lauf[48] bring our knowledge up-to-date regarding various types of antibody injury to the erythrocyte membrane and they consider the consequences thereof. Certainly, the antibody-injured red cell provides a fascinating model for future study.

Figure 1 summarizes some of the pathophysiologic pathways by which a normal deformable biconcave disc may be converted to a rigid erythrocyte. Those depicted in the upper half of the diagram include examples of rigid poikilocytes produced by virtue of changes in their contents, e.g., (A) sickle cells (B) Heinz body-containing cells, as well as cells damaged by extrinsic mechanical injury (C) which would include the hemolytic states associated with heart valve prostheses, as well as microangiopathic damage seen in small vessel disease such as thrombotic thrombocytopenic purpura or metastatic carcinoma. In the bottom half of the diagram, extrinsic damage by antibody (D) which may lead to partial phagocytosis[1,30,40] and the ATP dependent disc-sphere transformation (E) are illustrated as pathways leading to rigid spherocytic red cells. As will be pointed out by Dr. LaCelle,[29] although the hereditary spherocyte may have normal ATP levels, it undergoes the disc-sphere

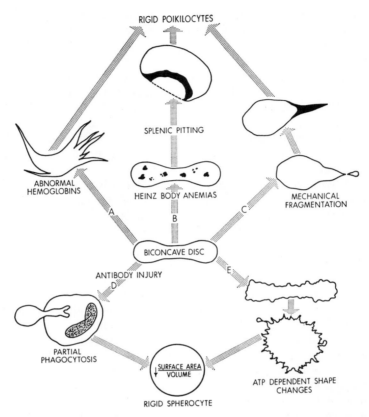

Fig. 1.—Erythrocyte fragmentation and rigidity. (Reproduced with permission from Plenary Session Papers, XII Congress International Society of Hematology, New York, 1968.)

shape changes as well as changes in intrinsic membrane rigidity at normal levels of ATP. Figure 1 provides a scheme by which one can relate abnormalities in red cell glycolysis, hemoglobin properties, antibody injury, loss of cellular deformability and altered morphology, including evidence of fragmentation.

A second major purpose of this issue of SEMINARS is to present an up-to-date evaluation of the ways in which the chemistry and structure of the membrane can be studied. The reviews by Drs. Cooper,[5] Maddy[31] and Weinstein and McNutt[65] consider three different ways by which the membrane can be analyzed. Dr. Cooper has dealt with the composition and turnover of red cell membrane lipids as well as summarizing abnormalities which have been reported in acquired and inherited membrane lipid abnormalities. He appropriately warns us, however, that we must still understand more about the architecture and physiology of the membrane before we can go directly from lipid analyses to an understanding of the relationship between a defect in abnormal cells and their attendant properties. For example, although the increased lipid and decreased osmotic fragility seen in obstructive jaundice represent a predictable relationship, the reason for the decreased filterability of such cells and their tendency to undergo spur formation are not clear from the properties of the lipids themselves.

By comparison to the documentation of membrane lipid abnormalities in disease states, only a few reports of membrane protein abnormalities exist to date.[20,21,35,36] Dr. Maddy[31] has aptly summarized the analytical difficulties which underlie the limited amount of work done in this field. In this review, Maddy has provided a realistic appraisal of technics which are available for study of abnormalities of red cell membrane protein and the information should be a most valuable guide for interested workers. This particular field very likely will ultimately prove to be the most rewarding, as well as the most challenging, approach for understanding normal and diseased membranes.

Although not included in a separate review, the importance and relevance of membrane carbohydrates are mentioned in Maddy's review of membrane protein,[31] Cooper's review of membrane lipids,[5] and are also mentioned in regard to the carbohydrate nature of various specific erythrocyte membrane antigens by Rosse and Lauf.[48] Readers interested in a more detailed discussion directed primarily toward erythrocyte membrane carbohydrates per se should consult the review by Rapport.[45]

Drs. Weinstein and McNutt[65] have brought us an up-to-date review of the exciting new electron microscopic freeze-cleave technic which promises to contribute greatly to our future understanding of normal and pathologic red cell membranes. Weinstein and McNutt also point out that much of the electron microscopic work on abnormal erythrocytes done in the past is subject to serious technical criticisms, including many fixation artifacts and consequently, cannot be accepted as definitive. We are still regrettably far from the level of resolution which will permit us to relate our chemical studies directly to structural elements. Nevertheless, the electron microscopic approach still holds out exciting promise for hematologists who are morphologists at heart.

It clearly is no longer appropriate to consider separately disorders of hemoglobin structure or synthesis, metabolic abnormalities secondary to glycolytic enzyme deficiencies and "primary membrane disorders." The evolution in orientation toward consideration of interrelationships between these parts of the red cell is illustrated quite well by the fact that we ourselves in 1962 wrote a paper entitled "Is hemoglobin an essential structural component of erythrocyte membranes?"[57] and answered "no." More recently,[62] we have come to the conclusion that depending upon one's definition of "the membrane," the boundary between membrane and soluble interior of the cell may be quite dynamic and dependent upon cellular metabolism. In fact, what may be considered membrane may include some insoluble hemoglobin in ATP-depleted cells or as pointed out by Jacob,[22] the hemoglobin may be associated with the membrane as Heinz bodies. The story of the red cell membrane in disease states is unfolding at an ever quickening pace, and although much remains to be learned at the molecular level, it is truly encouraging to be able to relate a variety of disorders through the final common pathway of cellular deformability.

REFERENCES

1. Archer, G. T.: Phagocytosis by monocytes of red cells coated with Rh antibodies. Vox Sang. 10:590, 1965.

2. Bessis, M., Bricka, M., and Dupuy, A.: Examen au microscope électronique de la surface des globules rouges. Origine des hématexodies. C. R. Soc. Biol. (Paris) 145:1509, 1951.

3. Brown, A.: Morphological changes in the red cells in relation to severe burns. J. Path. Bact. 58:367, 1946.

4. Carson, P. E., Flanagan, C. L., Ickes, C. E., and Alving, A. S.: Enzymatic deficiency in primaquine-sensitive erythrocytes. Science 124:484,1956.

5. Cooper, R. A.: Lipids of human red cell membrane: Normal composition and variability in disease. Seminars Hemat. 7:296, 1970.

6. Dacie, J. V.: The Haemolytic Anaemias, Congenital and Acquired. London, Churchill, 1954, p. 366.

7. —, Mollison, P. L., Richardson, N., Selwyn, J. G., and Shapiro, L.: Atypical congenital haemolytic anaemia. Quart. J. Med. 22:79, 1953.

8. Dameshek, W., and others: Case records of the Massachusetts General Hospital. Case 52, 1964, New Eng. J. Med. 271:898, 1964.

9. DeSandre, G., Ghiotto, G., and Mastella, G.: L'acetilcolinesterasi eritrocitaria. II. Rapporti con le mulattie emolitiche. Acta Med. Patav. 16:310, 1956.

10. Dodge, J. T., Mitchell, C., and Hanahan, D. J.: The preparation and chemical characteristics of hemoglobin-free ghosts of human erythrocytes. Arch. Biochem. 100: 119, 1963.

11. Ehrlich, P.: Farbenanalytische Untersuchungen zur Histologie and Klinik des Blutes. Berlin, 1891, p. 99.

12. Fessas, P., Loukopoulos, D., and Thorell, B.: Absorption spectra of inclusion bodies in beta thalassemia. Blood 25:105, 1965.

13. Fuhrmann, G. F.: Beeinflussung der Kalium und Natriumpermeabilität menschlicher Erythrocyten durch Harnstoff. Arch. Ges. Physiol. 283:R66, 1965.

14. Firkin, B. G., and Wiley, J. S.: The red cell membrane and its disorders. Prog. Hemat. 5:26, 1966.

15. Ganzoni, A., Hillman, R. S., and Finch, C. A.: Maturation of the macroreticulocyte. Brit. J. Haemat. 16:119, 1969.

16. Gasser, W. C., Gautier, E., Steck, A., Siebenmann, R. E., and Dechslin, R.: Hämolytisch-urämische syndrome: bilaterale Nierenrindennekorsen bei akuten erworbenen Hämolytischen Anämein. Schweiz. Med. Wschr. 85:905, 1955.

17. Ginn, F. L., Hochstein, P., and Trump, B. F.: Membrane alterations in hemolysis: internalization of plasmalemma induced by primaquine. Science 164:843, 1969.

18. Ham, T. H., Shen, S. C., Fleming, E. M., and Castle, W. B.: Studies on the destruction of red blood cells. IV. Thermal injury: action of heat in causing increased spheroidicity, osmotic and mechanical fragilities and hemolysis of erythrocytes; observations on the mechanism of destruction of such erythrocytes in dogs and in a patient with a fatal thermal burn. Blood 3:373, 1948.

19. Haradin, A. R., Weed, R. I., and Reed, C. F.: Changes in physical properties of stored erythrocytes. Transfusion 9:229, 1969.

20. Harvald, B., Hanel, K. H., Squires, R., and Trap-Jensen, J.: Adenosine-triphosphatase deficiency in patients with nonspherocytic haemolytic anaemia. Lancet 2:18, 1964.

21. Hartmann, R. C., and Auditore, J. V.: Paroxysmal nocturnal hemoglobinuria. I. Clinical studies. II. Erythrocyte acetylcholinesterase defect. Amer. J. Med. 27:398, 1959.

22. Jacob, H. S.: Mechanisms of Heinz body formation and attachment to the red cell membrane. Seminars Hemat. 7:341, 1970.

23. Jandl, H. J., Simmons, R. L., and Castle, W. B.: Red cell filtration in the pathogenesis of certain hemolytic anemias. Blood 28:133, 1961.

24. Jandl, J. H.: Leaky red cells. Blood 26:367, 1965.

25. —, Engle, L. K., and Allen, D. W.:

Oxidative hemolysis and precipitation of hemoglobin. I. Heinz body anemias as an acceleration of red cell aging. J. Clin. Invest. 39:1818, 1960.

26. Jensen, W. N., Bromberg, P. A., and Bessis, M. C.: Micro-incision of sickled erythrocytes by a laser beam. Science 155: 704, 1967.

27. —, and Lessin, L.: Morphologic abnormalities associated with hemoglobinopathies. Seminars Hemat. October 1970.

28. Koyama, S., Aoki, S., and Deguchi, K.: Electron microscopic observations of the splenic red pulp with special reference to the pitting function. Mie Med. J. 14:143, 1964.

29. LaCelle, P. L.: Alteration of membrane deformability in hemolytic anemias. Seminars Hemat. 7: October, 1970.

30. Lobuglio, A. F., Cotran, R. S., and Jandl, J. H.: Red cells coated with immunoglobulin G.: binding and sphering by mononuclear cells in man. Science 158:1582, 1967.

31. Maddy, A. H.: Erythrocyte membrane proteins. Seminars Hemat. 7:275, 1970.

32. Marmont, A., and Bianchi, V.: Mediterranean anemia. Clinical and haematological findings and pathogenetic studies in the milder form of the disease. Acta Haemat. 1:4, 1948.

33. Meltzer, S. J., and Welch, W. H.: The behavior of the red blood corpuscles when shaken with indifferent substances. J. Physiol. (London) 5:255, 1884.

34. Mitchison, J. M., and Swann, M. M.: The mechanical properties of the cell surface. I. The cell elastimeter. J. Exp. Biol. 31:443, 1954.

35. Mircevová, L., Brabec, V., and Palek, J.: Adenosine-triphosphatase activity in hereditary spherocytosis. Blut 15:1, 1967.

36. Nakao, K., Kurashina, S., and Nakao, M.: Adenosine-triphosphatase activity of erythrocyte membrane in hereditary spherocytosis. Life Sci. 6:595, 1967.

37. Nakao, M., Nakao, T., and Yamazoe, S.: Adenosine triphosphatase and maintenance of shape of the human red cells. Nature 187:945, 1961.

38. Nathan, D. G., and Shohet, S. B.: Erythrocyte ion transport defects and hemolytic anemia: "hydrocytosis" and "desiccytosis." Seminars Hemat. 7: October, 1970.

38a. Neerhout, R. C.: Disorders of the red cell membrane: A review of biochemical and physiologic alterations of erythrocyte membranes which may lead to morphologic changes and shortened cell survival. Clin. Ped. 7:451, 1968.

39. Pauling, L., Itano, H. A., Singer, S. J., and Wells, I. C.: Sickle cell anemia: a molecular disease. Science 110:543, 1949.

40. Policard, A., and Bessis, M.: Fractionnement d'hématies par les leucocytes au course de la phagocytose. C. R. Soc. Biol. (Paris) 147:982, 1953.

41. Ponder, S.: The prolytic loss of K+ from human red cells. J. Gen. Physiol. 30:235, 1947.

42. Ponder, E.: Hemolysis and Related Phenomena. New York, Grune & Stratton, 1948.

43. Rand, R. P., and Burton, A. C.: Mechanical properties of the red cell membrane. I. Membrane stiffness and intracellular pressure. Biophys. J. 4:115, 1964.

44. —: Mechanical properties of the red cell membrane. II. Viscoelastic breakdown of the membrane. Biophys. J. 4:303, 1964.

45. Rapport, M. M.: Some chemical aspects of membrane structure. Plenary Session Papers. XII Cong. Int. Soc. Hemat., New York, 1968, p. 76.

46. Reed, C. F., and Swisher, S. N.: Erythrocyte lipid loss in hereditary spherocytosis. J. Clin. Invest. 45:777, 1966.

47. Rose, J. C., Hufnagel, C. A., Fries, E. D., Hariey, W. P., and Partenope, E. A.: The hemodynamic alterations produced by a plastic valvular prosthesis for severe aortic insufficiency in man. J. Clin. Invest. 33:891, 1954.

48. Rosse, W. F., and Lauf, P. K.: Effects of immune reactions on the red cell membrane. Seminars Hemat. 7:323, 1970.

49. Rous, P.: Destruction of the red blood corpuscles in health and disease. Physiol. Rev. 3:75, 1923.

50. Schultze, M.: Ein heizberger Objekttisch und seine Verwendung bei Untersuchung des Blutes. Arch. Mikr. Anat. 1:1, 1865.

51. Sears, D. A., and Crosby, W. H.: Intravascular hemolysis due to intracardiac prosthetic devices: diurnal variation related to activity. Amer. J. Med. 39:341, 1965.

52. Slater, L. M., Muir, W. A., and Weed, R. I.: Influence of splenectomy on insoluble hemoglobin inclusion bodies in β thalassemic erythrocytes. Blood 31:766, 1968.

53. Teitel, A., and Rodulesco, I.: O

metoda de determinaire a plasticitatii globu-
lelor rosii. Med. Interna. (Bucharest) 5:32,
1952.

54. Teitel, P.: Le test de la filtrabilité
erythrocytaire (TFE) une methode simple
d'étude de certaines proprietes microrheo-
logique des globules rouges. Nouv. Rev.
Franc. Hemat. 7:195, 1967.

55. Tosteson, D. C.: Potassium exchange
in sickle cell anemia red cells. J. Clin.
Invest. 32:608, 1951.

56. Weed, R. I., LaCelle, P. L., and Merrill, E. W.: Metabolic dependence of red
cell deformability. J. Clin. Invest. 48:795,
1969.

57. —, Reed, C. F., and Berg, G.: Is
hemoglobin an essential structural component of the human erythrocyte membrane?
J. Clin. Invest. 42:581, 1963.

58. —, and —: Membrane alterations leading to red cell destruction. Amer. J. Med.
41:681, 1966.

59. —, and Bowdler, A. J.: Metabolic
dependence of the critical hemolytic volume
of human erythrocytes. Relationship to osmotic fragility and autohemolysis in hereditary spherocytosis and normal red cells. J.
Clin. Invest. 46:1137, 1966.

60.—, and Weiss, L.: The relationship of
red cell fragmentation occurring within the
spleen to cell destruction. Trans. Ass. Amer.
Physicians 79:426, 1966.

61. —: The cell membrane in hemolytic
disorders. Plenary Session Papers, XII Cong.
Int. Soc. Hemat., New York, 1968, p. 81.

62. Weiss, L., and Tavassoli, M.: Anatom-
ical hazards to the passage of erythrocytes
through the spleen. Seminars Hemat. 7:
October, 1970.

63. Wennberg, E., and Weiss, L.: Splenic
erythroclasia: An electron microscopic study
of hemoglobin H. disease. Blood 31:778,
1968.

64. Weinstein, R. S., and McNutt, N. S.:
The ultrastructure of red cell membranes.
Seminars Hemat. 7:259, 1970.

65. Westerman, M. P., Pierce, L. E., and
Jensen, W. N.: Erythrocyte lipids: a comparison of normal young and normal old
populations. J. Lab. Clin. Med. 62:394,
1963.

Ultrastructure of Red Cell Membranes

By Ronald S. Weinstein and N. Scott McNutt

T HREE DECADES of electron microscopy of red cells have confirmed
that the red cell has a membrane near or at its surface and that this
membrane is very thin. Aside from these points, the microscopists' contributions to red cell membraneology have been largely equivocal. Several categories
of problems might be approachable with the electron microscope if adequate
preparative technics were available. These include: testing hypotheses (e.g.,
membrane models) for the general molecular architecture of membranes;
characterizing lesions in abnormal membranes; identifying sites for membrane
transport; and correlating membrane structure with the functional state. While
these are important areas for investigation, the severe limitations imposed by the
preparative technics are frequently impassable obstacles that prevent the
microscopist from extracting definitive information from the optical images
he so reproducibly obtains with his microscope. Since the appearance of
membranes at high resolution is directly related to the methods used to
obtain the pictures, a meaningful discussion of red cell membrane ultrastructure must take into account some artifacts produced by the preparative
technics. In this short summary, we will outline briefly the appearance of red
cell membranes as seen with the commonly used preparative technics. These
appearances will be related to some of the broad categories of problems just
outlined.

Thin-Sectioned Membranes

When red cells are fixed in osmic acid (OsO_4), embedded in plastic and
thin sectioned, the dense osmium reaction product at the perimeter of the
cell appears as a triple-layered structure corresponding in location to the
cell membrane. The membrane's molecules contribute little to the image.
As seen in the electron microscope, the triple-layered structure consists of
two outer dense lines, each 25Å in thickness, separated by a 20 to 30Å central
electron lucent zone. This complex accurately reflects the location of the main
barrier to free diffusion at the cell periphery.

The triple-layered appearance has been termed the "unit membrane" by
Robertson[61] and is widely used as evidence for the lipid bilayer model for
membrane structure proposed by Danielli and Davson.[11] This model shows
an extended bimolecular lipid leaflet sandwiched between layers of nonlipids,

---

*From the Department of Pathology and Neurosurgical Service, Massachusetts General
Hospital and Department of Pathology, Harvard Medical School, Boston, Mass.*

Ronald S. Weinstein, M.D.: *Head, Mixter Laboratory for Electron Microscopy; and
Clinical and Research Fellow, Massachusetts General Hospital, Boston, Mass.; and Teaching
Fellow, Harvard Medical School, Cambridge, Mass.* N. Scott McNutt, M.D.: *Clinical and
Research Fellow, Department of Pathology, Massachusetts General Hospital, Boston, Mass.;
and Teaching Fellow, Harvard Medical School, Cambridge, Mass.*

mainly proteins. The lipid molecules are radially oriented with the polar groups facing outward and the long-chain hydrocarbons with terminal $-CH_3$ groups pointing inward. Hydrophobic bonding predominates in the interior of the membrane. Adsorbed non-lipid layers maintain their position, at least in part, by ionic bonding to the underlying hydrophilic head groups of the membrane phospholipids. Robertson has interpreted the two dense lines seen in thin sections of membranes as representing the polar head groups of membrane phospholipids stained with osmic acid, while the central lucent zone represents unstained long hydrocarbon chains of lipid molecules extending into the interior of the membrane.[61]

The specificity of the reaction of lipids with osmic acid (OsO_4) is now in doubt since OsO_4 apparently reacts with double bonds of unsaturated lipids but not with polar head groups.[39-42] Simultaneously, an approximately equivalent amount of OsO_4 is converted to osmium dioxide (OsO_2). The OsO_2 may orient at the aqueous-organic interface of membranes in a nonspecific fashion that in no way reflects the presence or absence of polar head groups near the interface.[41,42] Other observations enrich the argument that the triple-layered appearance is not specific for membranes containing lipid bilayers. For example, the triple-layered appearance is retained in membranes of lipid-extracted mitochondria[21] and the periodicity of the wrapping of myelin membranes of nerve survives lipid extraction.[59] Further, pure protein membranes in some bacterial cells also appear triple-layered in thin sections.[68] It is worth stressing that, although the triple-layered appearance of thin sectioned membranes does not argue strongly for or against the Danielli model, a large body of indirect evidence suggests that a lipid bilayer may exist in at least part of the red cell membrane. (c.f.[67])

In sections, membranes appear as monotonous structures with no evidence of focal specialization. Interestingly, ferritin labeling[44] and histochemical technics[49] have demonstrated in thin sections that blood group antigens and some enzymes have characteristic patterns of distribution on the membrane, although the underlying membranes appear undifferentiated in the sections.

In ordinary sections, surface views of unstained membranes are particularly unrevealing since the membranes are virtually electron lucent. Even though globular subunits occasionally have been described in *en face* views of a few kinds of stained sectioned membranes (not red cells), the location of the subunits within the thickness of the membrane cannot be determined on the basis of their electron microscopic image alone.[62]

Negative Stained Membranes

The surface topography of red cell ghost membranes has been investigated by the negative staining technic.[9,25] This method requires that ghost membranes* and a solution of heavy metal salt be dried together onto thin supporting films. In such preparations, structures that are impermeable to

---

*The negative staining technic cannot be used for studying the membranes of intact red cells.

the heavy metal appear electron lucent and regions permeated by stain appear electron dense. The subunits seen on a few kinds of membranes with negative staining are inapparent on red cell ghosts[9] which can appear quite structureless when stained at room temperature.[25] Staining at 37°C demonstrates a 90Å particulate component associated with the ghost membranes[25] but, as is the case with thin sections, the location of the subunits on or within the membrane cannot be determined.

Some red cell membrane proteins have been isolated from ghost preparations and examined with negative staining.[27,28,50,51] These studies tell nothing about the precise position or function of the molecules before they are extracted or even if they are an essential membrane structural component.[28,51] The negative staining technic has been useful for identifying lesions in pathologically altered membranes, as will be described later.

Air-Dried/Metal-Shadowed Membranes

Some surfaces of red cell ghost membranes can also be examined by drying ghosts from an aqueous solution onto thin supporting films and then shadowing their topography by condensing metal vapor onto them in vacuo. The final electron microscopic image is determined by variations in thickness of the electron-dense shadowing material caused by irregularities in the membrane's surface topography, while the contribution of the underlying membrane itself to the image is negligible. Of the many studies on such preparations (as reviewed by Ponder[60]) those of Hillier and Hoffman[30,31,33] are most frequently quoted. Their preparations have shown that the major portion of the membrane appears to be composed of what they believe to be short cylinders or "plaques" about 30Å thick and from 100 to 500Å in diameter.[30] These plaques rest on an incomplete network of 20Å thick fibers oriented tangentially to the surface of the cell.

Such observations have been used to support "subunit" models of membrane structure. Several different models have been proposed based on lipoprotein or lipid-protein subunits aggregated into continuous thin sheets or lattices.[2,23,24,30,36,69] It is conceivable that sheets of subunits, viewed *en face,* could give the appearance described for air-dried/metal-shadowed preparations. However, the argument loses much of its appeal when the artifacts and limitations of the air-dried/metal-shadow technic are considered.

Since membrane cross-sections are not seen in these preparations, the relation of the surface topography viewed *en face* to the membrane's internal structure is obscure. A serious criticism of the technic results from the microscopist's inability to attribute the observed structures to the membrane itself, rather than to "extraneous" material on the membrane surface.[70] Material of functional importance to the membrane may be located outside the limits of its major permeability barrier.[1] By convention, microscopists consider material outside the boundaries marked by the dense lines of the triple-layered membrane seen in sections as "extraneous coat." Extensive washing of red cell ghosts may remove some extraneous material, such as hemoglobin and absorbed plasma constituents. However, there is no way to be certain that all "extraneous" material is removed prior to air-drying. When shadowed, any

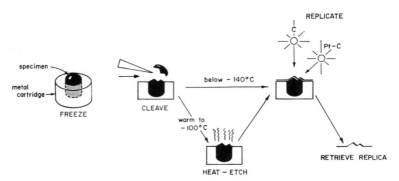

Fig. 1.—The freeze-cleave protocol consists of the following steps: 1. A pellet of packed cells or ghosts is frozen rapidly in Freon at −150°C. 2. The frozen specimen is cleaved at low temperature with a razor blade. 3. The newly produced fracture face is replicated at low temperature by evaporating platinum and carbon onto it in a vacuum system. The condensing platinum-carbon mixture forms a coherent replica of the topography of the fracture face. 4. After thawing of the specimen, the replica is retrieved, picked up on grids and examined in a standard electron microscope. An optional step, heat-etching, can be added to the protocol to demonstrate some structures that are not seen with freeze-cleaving alone.

remaining "extraneous" material could contribute to the membranes' topography and thus to the image produced after metal-shadowing.

A discussion of other evidence favoring the idea that sheets of subunits form membranes is beyond the scope of this presentation and has recently been reviewed elsewhere by Stoeckenius and Engelman[67] and Korn.[42] Suffice it to say that, in our opinion, there is no compelling physical evidence that the red cell membrane is constructed of sheets of small subunits.

Freeze-Cleaved Membranes

A major criticism of all three technics discussed thus far centers on the necessity of removing membrane water, since water may be an essential structural component and its removal may induce significant alterations.[20,29,36,57] Newer preparative methods have been devised with the goal of preserving membranes in a hydrated state. Freeze-cleaving is one such technic that circumvents many of the artifact-producing steps of other technics[37,55,56,74] and has been applied to the problem of red cell membrane ultrastructure in many laboratories.[18,24,34,38,51,54,70-75] Among the technic's advantages are its capacity for maintaining membranes in a partially hydrated state and its ability to demonstrate large areas of membrane in remarkable three-dimensional relief at relatively high resolution.* Since freeze-cleaving has not as yet found its way into the standard electron microscopy textbooks, we will describe briefly the technic before proceeding to a discussion of the image of red cells obtained.

The freeze-cleave technic is outlined in Fig. 1. Packed red blood cells or

---

*Large areas of membrane surface can also be viewed with the stereoscan electron microscope, but the resolution currently attainable is relatively poor compared to that routinely achieved with replicas of freeze-cleaved red cells.

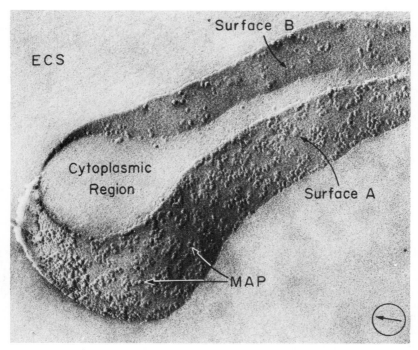

Fig. 2.—Replica of a freeze-cleaved human red cell ghost membrane. The membrane appears as a thin sheet. Its A-Surface is oriented toward the extracellular space *(ECS)* and is partially covered with clusters of protruding 100Å membrane-associated particles (MAP). Surface-B has fewer particles and faces the cell's interior. The cytoplasmic region of this hemoglobin-free ghost appears empty. The encircled arrow shows the direction of platinum shadowing. This preparation is not heat etched. 120,000× (Weinstein and Koo,[73] with permission of Academic Press, Inc., labeling has been updated.)

red cell ghosts are rapidly frozen in a small metal cartridge. The frozen specimen is fractured either in liquid nitrogen (−196°C) or in vacuo with a pre-cooled razor blade. As the cleavage passes through the frozen specimen, it follows along natural paths of low mechanical resistance within the specimen. These paths are frequently along membranes. Large areas of membrane are thus exposed at the fracture faces produced by cleaving. The faces are replicated by condensing a mixture of vaporized carbon and platinum onto them in a vacuum system. The frozen cells or membrane ghosts are then thawed and the cells are digested from the replicas. The replicas are washed, placed on grids and examined in a transmission electron microscope. An additional step, called "heat-etching", may be added to the protocol. Its value will become apparent later.*

---

*The freeze-cleave technic is widely referred to as "freeze-etching," a name which is somewhat misleading. The "etch" step is an ancillary procedure that is useful in some applications but is often unnecessary. The essential steps are tissue freezing, cleaving, and replicating. The term "freeze-cleaving," therefore, is more suitable as a general term and etching should only be mentioned when it is utilized.

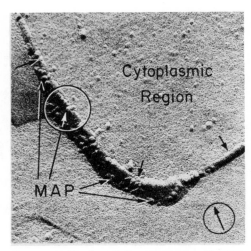

Fig. 3.—Replica of a partially lysed human red cell. The cleavage plane has exposed a small area of membrane (an A-Surface), covered with particles *(MAP)*, before passing across the leaflet (short arrows) into the cytoplasmic region. One particle (encircled) appears cylindrical and extends through the membrane leaflet to its juxtacytoplasmic surface. Such particles may represent specialized sites in membranes where transport occurs. 120,000 ×. (Weinstein and Koo,[73] with permission of Academic Press, Inc.).

In preparations of freeze-cleaved red cells, two distinctive surfaces of red cell membranes are visualized.[5,34,38,54,70,75] Both surfaces are finely granular but are distinguished from one another by the presence of particles, 100 Å in diameter, that are four to five times as plentiful on one surface (Surface A) as on the other surface (Surface B) (Fig. 2). The surface covered with the greater number of particles is oriented toward the extracellular compartment and the surface with the lesser number of particles is oriented toward the cell's cytoplasmic compartment.[5,54,70] Some of these membrane-associated particles apparently penetrate through the entire thickness of the thin, sheet-like membrane[72,73] (Fig. 3), raising the possibility that they represent specialized sites in the membrane.[70]

Counts of membrane-associated particles on human red cell fracture faces have ranged from 2600[70] to 3800[54] per μ^2 of surface area for the A-Surface and from 575[70] to 1400[54] per μ^2 for the B-Surface. Assuming a surface area of 145 μ^2 for a normal human red cell,[76] there is a total of 4.6 x 10^5 to 9.6 x 10^5 particles per cell. Particle counts are remarkably constant from cell to cell and donor to donor.[70] The particles survive membrane ghosting[73] and vesiculation in hypotonic buffers,[65] and serve as a useful marker of membrane orientation.[70]

"Heat-etching" demonstrates another surface of red cell membranes that is not seen with freeze-cleaving alone. When specimens are cleaved at temperatures below −140°C, the specimen ice is virtually nonvolatile. If the specimen temperature is raised after cleaving, the water present as small ice crystals in the specimen begins to sublime away. Hydrated regions in

Fig. 4.—Additional membrane surfaces are revealed by heat - etching When a red cell is freeze - cleaved (A), large areas of an aspect of its membrane can be exposed for replication. With heat-etching (B), ice surrounding the cell sublimes away and, as the fracture face recedes, additional areas of cell surface are exposed for replication. The level at which the initial cleavage was deviated onto the membrane (X→) is s h a r p l y defined. Typically, a thin layer of material is revealed by etching. (Weinstein,[70] with permission of J. B. Lippincott, Co.).

the fracture face recede as the ice sublimes away and poorly hydrated regions are brought out in three dimensional relief. This process is illustrated schematically in Fig. 4. A fracture passes through frozen water in which a red blood cell is suspended and is deviated onto the cell's membrane (Fig. 4, top). With heat-etching (Fig. 4, bottom), the water in which the cell is suspended sublimes away. As the water-ice recedes, a surface of the cell is revealed which previously was unseen. For red cells, the appearance of this layer is related to the composition of the suspending medium. When cells are frozen directly in plasma, prolonged heat-etching reveals a thin, textured coating layer (Fig. 5) with focal thickenings that may represent adsorbed plasma protein.[70] When washed cells are frozen in distilled water, the layer is uniformly thin, measuring 30 to 40 Å in thickness, and appears finely granular or smooth.

Initial reports suggested that the two surfaces revealed by cleaving represented the true outer and inner (juxtacytoplasmic) surfaces of the membrane.[34,38,70-75] The thin layer revealed by heat-etching was interpreted as representing an extraneous cell coat[70] similar to that on the surfaces of some avian red cells.[15,19] These interpretations were consistent with those made by many other investigators at the time on a large spectrum of membranes other than those of red cells (see references 37 and 55). Recent studies on the paths of cleavages along membranes now show that a reinterpretation of the earlier reports is necessary.

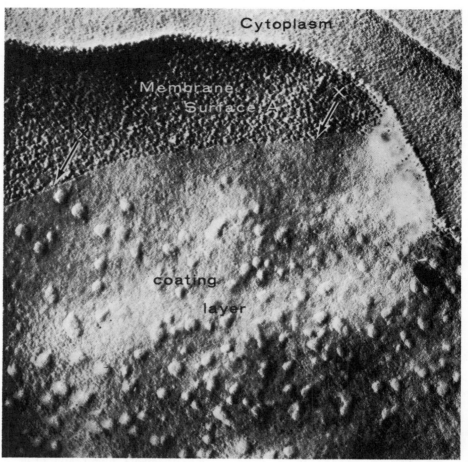

Fig. 5.—Replica of a freeze-cleaved and heat-etched human red cell. The level at which the cleavage plane originally deviated (see Fig. 4) is sharply defined (X→). After passing along the membrane Surface-A, the cleavage breaks into the cell's cytoplasm which appears coarsely granular in this intact cell. Heat-etching has revealed a thin layer which includes the membrane's outer leaflet and the tenuous "extraneous" coat. The cell was frozen directly in plasma. The focal thickenings of the coating layer may represent adsorbed plasma proteins. 56,000X (Weinstein and Someda,[70] with permission of J. B. Lippincott, Co. Labeling has been updated).

In 1966, Branton proposed that cleavages, rather than passing along the true outer and inner surfaces of membranes, actually pass through the interior of membranes and that the two surfaces demonstrated by cleaving represent previously unseen faces that originate from within the membrane.[3] He interpreted the thin coating layer revealed by heat-etching as representing half the thickness of the membrane. According to the Branton interpretation (see Fig. 6), Surface-A and Surface-B face each other inside the membrane, rather than facing in opposite directions as previously assumed. Membrane-

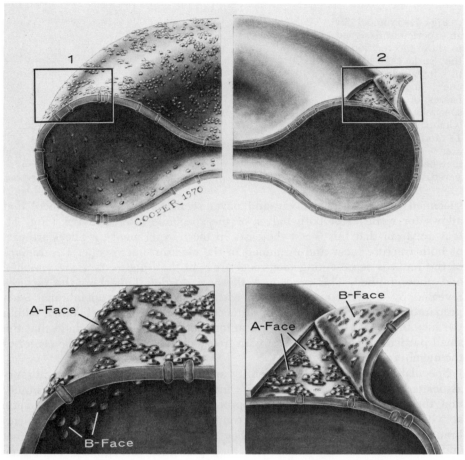

Fig. 6.—Artist's concept of the alternative interpretations of the location of the A-Surface and B-Surface of red cell membranes demonstrated with freeze-cleaving. Interpretation 1 shows the membrane-associated particles on the true outer (Face-A) and inner (Face-B) surfaces of the membrane. The two surfaces face in opposite directions. Interpretation 2 shows the fracture entering the membrane and passing through its interior. Face-A and Face-B point toward each other within the membrane.

associated particles could still extend through the full thickness of the membrane. Deamer and Branton provided conclusive evidence of membrane splitting in artificial lipid membrane systems[14] and Engstrom and Branton suggested that the interpretation could be applied to red cell membranes as well.[18]

Direct evidence for the applicability of Branton's membrane splitting hypothesis to red cell membranes has recently been obtained in this laboratory. Normally, when a specimen is cleaved the part of the specimen that is cleaved away is lost (Fig. 1). Investigators who believe that the true outer surface of the membrane is revealed by cleaving assume that material next to the membrane (extracellular fluid or cytoplasm) is retained in the part of the

Fig. 7.—Protocol for an experiment designed to show the relation of the surfaces revealed by freeze-cleaving to one another. Both halves of the cleaved specimen are replicated (Compare with Fig. 1) and corresponding areas of the replicas matched and compared.

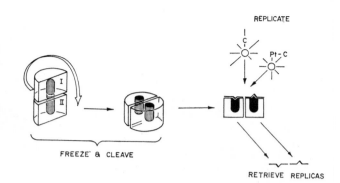

specimen that is cleaved away and never seen. Now, using a protocol that allows for retrieval of both halves of the specimen after cleaving (Fig. 7), it is apparent that this is not the case. In these experiments, replicas are cast of both fracture faces and examined in the electron microscope. The regions of the replicas representing reciprocal areas in the fracture face are matched and photographed. This procedure allows us to determine how the surfaces revealed by cleaving are related to each other within the membrane. Such preparations show that A and B surfaces do indeed face each other within the membrane,[71a] as predicted by Branton, and allow us to conclude that the small particulate component seen with freeze-cleaving exists near the center of the membrane (Fig. 6, right).

Speculation on the chemical identity and functional role of the membrane-associated particles in the membrane's interior has centered on the possibility that they are either proteins or lipoproteins,[5,54,55,58] possibly with enzymatic activity[55] and possibly involved in transporting materials across the membrane.[54,70,72] Current evidence that they are involved in transport is indirect and based on the observation that they penetrate through the membrane leaflets;[73] the fact that they represent a small percentage of the membrane's total surface area;[70,73] the absence of membrane-associated particles on membranes such as myelin[4,74] which are devoid of transport functions; and on their high density at specialized regions of some cell membranes [7,43,53] where transport is the major function and the barrier function is minimal. The possibility that they could represent lipid micelles[5] or some as yet unrecognized artifact of preparation is widely acknowledged.[7,37,71,74]

The existence of natural cleavage planes within membranes provides strong support for a lipid bilayer within at least part of the red cell membrane.[3-5,67] A cleavage might easily follow along a pathway lined with terminal $-CH_3$ groups of membrane hydrocarbon chains at or near the center of the membrane.[3] If the membrane-associated particles represent protein within the membrane, then the Danielli model would require some refinement, perhaps along the lines suggested by Danielli himself.[10] He suggests that the surface layers of protein may occasionally penetrate the membrane to form small channels containing hydrophilic pores. The modifications required do not detract from the central theme of his model, the lipid bilayer, and indeed, add to its attractiveness since compositional and functional specificity of different kinds of membranes may be more easily explained. Protein within the membrane could also explain the apparent disparity between the amount of lipid in red cell membranes and the somewhat greater amount that would be required to form a complete bimolecular leaflet of lipid around the cell.[32]

Comparative Ultrastructure

The electron microscopic appearances of red cell membranes examined with the different preparative technics are reasonably consistent with one another if we take into account the limitations and artifacts of the various methods. For example, the membrane "subunits" seen in air-dried/metal-shadowed preparations may represent surface patterning with no counterpart within the membrane since the relation of the membrane's surface to its interior cannot be determined with that technic. Smaller "subunits" have been described on some surfaces of freeze-cleaved membranes, but no counterparts to these are seen in cross-fractures of membranes,[70] although a few of the larger membrane-associated particles do appear to extend through the entire thickness of the membrane. We can, therefore, conclude that most of the surface patterning seen in flat areas of a membrane in air-dried or freeze-cleave preparations is consistent with the Danielli model since the patterning may not involve a significant thickness of the membrane.

The membrane-associated particles demonstrated with freeze-cleaving are inapparent in thin sections. Since drastic changes in membrane components are induced during the preparation of cells for thin sectioning,[46,57] the inability of the technic to demonstrate such structures is not bothersome. Membrane-associated particles are not usually seen with negative staining. This deficiency is expected since structures demonstrable with negative staining are often confined to membrane outer surfaces, while membrane-associated particles are primarily within the membrane. It would be interesting to know if the ability to demonstrate 90 Å particles in red cell ghosts negatively stained at elevated temperature, as described by Haggis,[25] could be related to the greater penetration of stain into the membrane's interior at the higher temperature.

Assuming now that the membrane is hemisected by cleaving, then most of the thin layer revealed by heat-etching must represent a split thickness of the membrane. The surface revealed by heat-etching (Fig. 4) of cells suspended in distilled water must be very near or at the true surface of the membrane. High resolution studies of this surface have not been reported. The freeze-cleave data further suggest that "extraneous" coats such as those described on some avian red cells[15,19] must be very tenuous on human red cells.

The concept of the red cell membrane that emerges from ultrastructural studies is that of a thin, sheet-like structure that is occasionally interrupted by a small particulate component that may extend through the thin sheet to its outer surface. The natural cleavage plane in the center of the membrane argues in favor of a bimolecular leaflet within most of the membrane. The concept of protein subunits (represented as the particulate component of freeze-cleave preparations) extending through the membrane has been expressed in a number of membrane models[10,45,54,64,66] that represent variations on the Danielli hypothesis.[10]

Available data on the interrelationships between lipids and proteins in membranes are fragmentary and must be greatly expanded before specific modifications of the Danielli model can be considered. Hopefully, a combined ultrastructural-biochemical approach will aid in the collection of such information and aid in the further elucidation of the position and role of specific components in the red cell membrane.

Ultrastructure of Abnormal Cell Membranes

The identification and interpretation of "membrane lesions" in red cells from patients with hematologic disorders is of great interest. Early efforts to detect membrane abnormalities involved using air-dried ghost membrane preparations. With this method, membrane lesions in two disease states were described. Hoffman and his associates found a coarse surface texture of red cell membranes in patients with thalassemia major.[33] Cells from patients with thalassemia minor and from normal donors appeared smoother. Danon et al., using similar methods, observed that red cells from patients with deficiencies in glucose-6-phosphate dehydrogenase appeared abnormally smooth,[13] similar to "old" cells of normal donors.[12] Since the relation of surface texture to the structure of the underlying membrane has not been defined, these observations tell us nothing about the derangement in the membrane of abnormal cells in physical and chemical terms.

Red cell membranes from patients with paroxysmal nocturnal hemoglobinuria (PNH) have been examined with several preparative technics. PNH is a rare hemolytic anemia that probably is related to an acquired red cell membrane defect. Some of these patients' red cells are apparently normal, while the remainder of the cells are abnormally sensitive to serum complement and tend to lyse in vivo.[26,63] In studying the membrane lesions in PNH, one may consider separately the initial acquired lesion that accounts for the membrane's increased sensitivity to complement and the lesions produced by complement. Efforts to visualize the initial acquired lesions have produced conflicting results with some workers claiming to find lesions[6,8,47,52] and others reporting that the red cells of these patients appear normal in the electron microscope.[16,48,75] As in the case with the abnormal textures of membranes described in thalassemia major and in glucose-6-phosphate dehydrogenase deficiency, the significance of "spots," "clefts," "irregular thickenings," "surface granularity" and "deep pits" described as the initial acquired lesions of PNH cells by various authors[6,8,47,52] is obscure.

In contradistinction, the lesions produced by serum complement seem to be better documented. Using the negative staining technic, membrane "holes" have been demonstrated after exposure to complement and antibody.[17,35] When the membranes are viewed en face, the lesions appear to consist of a central dark region, 100 Å in diameter, rimmed by a narrow (electron lucent) zone. Negative stain has presumably accumulated in the central dark region and has been excluded from the light-appearing perimeters. Where the membrane is seen on edge, the clear ring appears to project from the membrane surface, forming a hollow cylinder that fills with negative stain. Some evidence suggests that the complement-induced holes are in a lipid layer of the membrane. The holes disappear when membrane lipids are extracted from formalin-fixed membranes with chloroform-methanol.[35] Treatment with trypsin, to perturb membrane proteins, or with buffer at low pH to detach antibody and complement components from the membrane fails to remove the holes. On the basis of these experiments and other considerations, Humphrey and Dourmashkin suggest that what appear to be holes in the membrane with negative staining may represent a focal rearrangement of the predominantly lipid layers of membranes into small micelles.[35]

The problems in evaluating observations on abnormal membranes are threefold. First, all of the general problems of interpretation encountered in studying the ultrastructure of normal membranes reappear when abnormal membranes are examined. Second, our lack of detailed information on the molecular organization of membranes emasculates any attempt to discuss membrane lesions in molecular terms. An adequate understanding of normal structure must serve as the backbone of any discussion of the pathogenesis of membrane lesions. And third, the problem of dealing with technical artifacts is magnified when abnormal membranes are examined. Abnormal membranes may have different sensitivities to the artifacts of preparation as compared with their normal counterpart. Differences in appearance could reflect different levels of membrane injury incurred during specimen preparation, rather than the appearance of the lesions as they exist in vivo. The microscopist has the formidable task of judging between these two possibilities. Resolution of this problem will require detailed physical and biochemical correlations with the morphology visualized.

REFERENCES

1. Bennett, H. S.: Morphological aspects of extracellular polysaccharides. J. Histochem. Cytochem. 11:14, 1963.

2. Benson, A. A.: On the orientation of lipids in chloroplast and cell membranes. J. Amer. Oil Chem. Soc. 43:265, 1966.

3. Branton, D.: Fracture faces of frozen membranes. Proc. Nat. Acad. Sci. U.S.A. 55:1048, 1966.

4. —: Fracture faces of frozen myelin. Exp. Cell Res. 45:203, 1967.

5. —: Membrane structure. Ann. Rev. Plant Physiol. 20:209, 1969.

6. Braunsteiner, H., Gisinger, E., and Pakesch, F.: Confirmation of a structural abnormality in the stroma of erythrocytes from paroxysmal nocturnal hemoglobinuria (PNH) after hemolysis in distilled water. Blood 11:753, 1956.

7. Bullivant, S.: Freeze-fracturing of biological materials. Micron 1:46, 1969.

8. Cecchi, E., and Conestabile, E.: Paroxysmal nocturnal haemoglobinuria. Electron-microscopic study of red blood-cells. Lancet 2:466, 1957.

9. Cunningham, W. P., and Crane, F. L.: Variation in membrane structure as revealed by negative staining technique. Exp. Cell Res. 44:31, 1966.

10. Danielli, J. F.: In: Danielli, J. F., Parkhurst, K. G. A., and Riddiford, A. C. (Eds.): Surface Phenomena in Chemistry and Biology. New York, Pergamon Press, 1958, p. 146.

11. —, and Davson, H.: A contribution to the theory of permeability of thin films. J. Cell. Comp. Physiol. 5:495, 1935.

12. Danon, D., and Marikovsky, Y.: Morphologie des membranes des erythrocytes jeunes et ages. Observations au microscope electronique. C. R. Soc. Biol. (Paris) 155:12, 1961.

13. —, Sheba, C., and Ramot, B.: The morphology of glucose 6 phosphate dehydrogenase deficient erythrocytes: Electron-microscopic studies. Blood 17:229, 1961.

14. Deamer, D. W., and Branton, D.: Fracture planes in an ice-bilayer model membrane system. Science 158:655, 1967.

15. Dickinson, P. C. T., Chang, T-W.,

and Weinstein, L.: Association of "lanthanum-staining material" with hemagglutination by Rubella virus. Proc. Soc. Exp. Biol. Med. 132:55, 1969.

16. Douglas, D., and Eaton, J. C.: Paroxysmal nocturnal haemoglobinuria: Marchiafava-Micheli syndrome. Lancet 1:946, 1955.

17. Dourmashkin, R. R., and Rosse, W. F.: Morphologic changes in the membranes of red blood cells undergoing hemolysis. Amer. J. Med. 41:699, 1966.

18. Engstrom, L., and Branton, D.: Observations on freeze-etched erythrocyte membranes. J. Cell Biol. 39:40a, 1968.

19. Fawcett, D. W.: Surface specializations on absorbing cells. J. Histochem. Cytochem. 13:75, 1965.

20. Fernández-Morán, H.: Forms of water in biologic systems and the organization of membranes. Ann. N.Y. Acad. Sci. 125:739, 1965.

21. Fleischer, S., Fleischer, B., and Stoeckenius, W.: Fine structure of lipid-depleted mitochondria. J. Cell Biol. 32:193, 1967.

22. Green, D. E., and Perdue, J. F.: Membranes as expressions of repeating units. Proc. Nat. Acad. Sci. U.S.A. 55:1925, 1966.

23. Gross, L.: Active membranes for active transport. J. Theor. Biol. 15:298, 1967.

24. Haggis, G. H.: Electron microscope replicas from the surface of a fracture through frozen cells. J. Biophys. Biochem. Cytol. 9:841, 1961.

25. Haggis, G. H.: Electron microscopic study of the disruption of red-cell membranes. Biophys. Biochim. Acta. 193:237, 1969.

26. Ham, T. H., and Dingle, J. H.: Studies on destruction of red blood cells. II. Chronic hemolytic anemia with paroxysmal nocturnal hemoglobinuria: Certain immunological aspects of the hemolytic mechanism with special reference to serum complement. J. Clin. Invest. 18:657, 1939.

27. Harris, J. R.: The isolation and purification of a macromolecular protein component from human red blood cell ghosts. Biochim. Biophys. Acta 188:31, 1969.

28. —: Some negative contrast staining features of a protein from erythrocyte ghosts. J. Molec. Biol. 46:329, 1969.

29. Hechter, O.: Role of water structure in the molecular organization of cell membranes. Fed. Proc. 24: (Suppl 15) s91, 1965.

30. Hillier, J., and Hoffman, J. F.: On the ultrastructure of the plasma membrane as determined by the electron microscope. J. Cell. Comp. Physiol. 42:203, 1953.

31. Hoffman, J. F.: On the reproducibility in the observed ultrastructure of the normal mammalian red cell plasma membrane. J. Cell. Comp. Physiol. 45:837, 1956.

32. —: Discussion. J. Gen. Physiol. 52 (Suppl.): s—185, 1968.

33. —, Wolman, I. J., Hillier, J., and Parpart, A. K.: Ultrastructure of erythrocyte membranes in thalassemia major and minor. Blood 11:946, 1956.

34. Huhn, V. D., and Grassmann, D.: Feinstruktur gefriergeätzter Erythrozytenschatten. Blut 18:211, 1969.

35. Humphrey, J. H., and Dourmashkin, R. R.: The lesions in cell membranes caused by complement. Advan. Immunol. 11:75, 1969.

36. Kavanau, J. L.: Structure and Function of Biological Membranes. San Francisco, Holden-Day, 1965, Vol. 1.

37. Koehler, J. K.: The technique and application of freeze-etching in ultrastructure research. In: Lawrence, J. H., Gofman, J. E., and Hayes, T. L. (Eds.): Advances in Biological and Medical Physics. New York, Academic Press, 1968, p. 1.

38. Koehler, J. K.: Freeze-etching observations on nucleated erythrocytes with special reference to the nuclear and plasma membranes. Z. Zellforsch. 85:1, 1968.

39. Korn, E. D.: Synthesis of bis-(methyl 9, 10-dihydroxystearate) osmate from methyl oleate and osmium tetroxide under conditions used for fixation of biological materials. Biochim. Biophys. Acta 116:317, 1966.

40. —: Modification of oleic acid during fixation of amoebae by osmium tetroxide. Biochim. Biophys. Acta 116:325, 1966.

41. —: A chromatographic and spectrophotometric study of the products of the reaction of osmium tetroxide with unsaturated lipids. J. Cell Biol. 34:627, 1967.

42. —: Cell membranes: structure and synthesis. Ann. Rev. Biochem. 38:263, 1969.

43. Kreutziger, G. O.: Freeze-etching of intercellular junctions in mouse liver. Proc. Electron Micro. Soc. Am. Meeting 26:234, 1968.

44. Lee, R. E., and Feldman, J. D.: Visualization of antigenic sites of human erythrocytes with ferritin-antibody conjugates. J. Cell Biol. 23:396, 1964.

45. Lenard, J., and Singer, S. J.: Protein

conformation in cell membrane preparations as studied by optical rotatory dispersion and circular dichroism. Proc. Nat. Acad. Sci. U.S.A. 56:1828, 1966.

46. —, and—: Alteration of the conformation of proteins in red blood cell membranes and in solution by fixatives used in electron microscopy. J. Cell Biol. 37:117, 1968.

47. Lewis, S. M., Danon, D., and Marikovsky, Y.: Electron-microscope studies of the red cell in paroxysmal nocturnal haemoglobinuria. Brit. J. Haemat. 11:689, 1965.

48. Malasková, V., Mach, O., Brabec, V., and Chrobák, L.: Elektronenoptisch nachweisbare Veränderungen am Stroma roter Blutkörperchen bie paroxysmaler nächtlicher Hämoglobinurie. Folia Haemat. (Leipzig) 82:328, 1964.

49. Marchesi, V. T., and Palade, G. E.: The localization of Mg-Na-K-activated adenosine triphosphatase on red cell ghost membranes. J. Cell Biol. 35:385, 1967.

50. —, and Steers, E., Jr.: Selective solubilization of a protein component of the red cell membrane. Science 159:203, 1968.

51. —, —, Tillack, T. W., and Marchesi, S. L.: Some properties of spectrin: A fibrous protein isolated from red cell membranes. In: Jamieson, G. A., and Greenwalt, T. J. (Eds.): Red Cell Membrane Structure and Function. Philadelphia, J. B. Lippincott, 1969, p. 117.

52. Matthes, M., Schubothe, H., and Lindemann, B.: Klinische und experimentelle Studien zur chronischen hämolytischen Anämie mit nächtlicher Hämoglobinurie (Typ Marchiafava-Micheli). Acta Haemat. 5:193, 1951.

53. McNutt, N. S., and Weinstein, R. S.: Interlocking subunit arrays forming nexus membranes. Biophysical Society Abstracts 14:101A, 1970.

54. Meyer, H. W., and Winkelmann, H.: Die Gefrierätzung und die Struktur biologischer Membranen. Protoplasma 68:253, 1969.

55. Moor, H.: Use of freeze-etching in the study of biological ultrastructure. Int. Rev. Exp. Path. 5:179, 1966.

56. — Mühlethaler, K., Waldner, H., and Frey-Wyssling, A.: A new freezing-ultramicrotome. J. Biophys. Biochem. Cytol. 10:1, 1961.

57. Moretz, R. C., Akers, C. K., and Parsons, D. F.: Use of small angle X-ray diffraction to investigate disordering of membranes

during preparation for electron microscopy. I. Osmium tetroxide and potassium permanganate. Biochim. Biophys. Acta 193:1, 1969.

58. Mühlethaler, K., Moor, H., and Szarkowski, J. W.: The ultrastructure of chloroplast lamellae. Planta 67:305, 1965.

59. Napolitano, L., Lebaron, F., and Scaletti, J.: Preservation of myelin lamellar structure in the absence of lipid. A correlated chemical and morphological study. J. Cell Biol. 34:817, 1967.

60. Ponder, E.: The cell membrane and its properties. In: Brachet, J., and Mirsky, A. (Eds.): The Cell. New York, Academic Press, 1961, Vol. II, p. 1.

61. Robertson, J. D.: The ultrastructure of cell membranes and their derivatives. Biochem. Soc. Sympos. 16:3, 1959.

62. —: The occurrence of a subunit pattern in the unit membranes of club endings in Mauthner cell synapses in goldfish brains. J. Cell Biol. 19:201, 1963.

63. Rosse, W. F., and Dacie, J. V.: Immune lysis of normal human and paroxysmal nocturnal hemoglobinuria (PNH) red blood cells. I. The sensitivity of PNH red cells to lysis by complement and specific antibody. J. Clin. Invest. 45:736, 1966.

64. Schmitt, F. O., and Davison, P. F.: Role of protein in neural function. Neurosci. Res. Prog. Bull. 3:55, 1965.

65. Steck, T. L., Weinstein, R. S., Straus, J. H., and Wallach, D. F. H.: Inside-out red cell membrane vesicles: preparation and purification. Science 168:255, 1970.

66. Stein, W. D.: Intra-protein interactions across a fluid membrane as a model for biological transport. J. Gen. Physiol. 54 (Suppl): S81, 1969.

67. Stoeckenius, W., and Engelman, D. M.: Current models for the structure of biological membranes. J. Cell Biol. 42:613, 1969.

68. —, and Kanau, W. H.: Further characterization of particulate fractions from lysed cell envelopes of Halobacterium Halobium and isolated gas vacuole membranes. J. Cell Biol. 38:337, 1968.

69. Wallach, D. F. H., and Gordon, A. S.: Lipid protein interactions in cellular membranes. Fed. Proc. 27:1263, 1968.

70. Weinstein, R. S.: Electron microscopy of surfaces of red cell membranes. In: Jamieson, G. A., and Greenwalt, T. J. (Eds.): Red Cell Membrane Structure and Function. Philadelphia, J. B. Lippincott, 1969, p. 36.

71. —, and Bullivant, S.: The application of freeze-cleaving technics to studies on red blood cell fine structure. Blood 29:780, 1967.

71a. —, Clowes, A. W., and McNutt, N. S.: Unique cleavage planes in frozen red cell membranes. Proc. Soc. Exp. Biol. Med. (in press).

72. —, and Koo, V. M.: Ultrastructure of freeze-cleaved and -etched red cell membranes and isolated red cell ghosts prepared by gradual osmotic lysis. J. Cell Biol. 35:190a, 1967.

73.—, and —: Penetration of red cell membranes by some membrane-associated particles. Proc. Soc. Exp. Biol. Med. 128: 353, 1968.

74. —, and Someda, K.: The freeze-cleave approach to the ultrastructure of frozen tissues. Cryobiology 4:116, 1967.

75. —, and Williams, R. A.: Freeze-cleaving of red cell membranes in paroxysmal nocturnal hemoglobinuria. Blood 30:785, 1967.

76. Westerman, M. P., Pierce, L. E., and Jensen, W. N.: A direct method for the quantitative measurement of red cell dimensions. J. Lab. Clin. Med. 57:819, 1961.

Erythrocyte Membrane Proteins

By A. H. MADDY

ANALYSIS OF THE ROLE of proteins in the structure and function of membranes may be pursued at three levels: the intact cell, the isolated membrane, and the individual molecular components. Most fruitful progress is achieved when all three levels are pursued concurrently, for, while each can make its unique contribution, each has its peculiar limitations.

Analysis of membrane activity in the living cell is confined to methods that do not damage the cell, yet, in spite of this severe restriction, permeability processes and the immunological properties of the surface have been extensively studied. This approach, however, usually does not permit a direct identification of the molecular nature of the active components. Thus, permeability measurements allow the construction of phenomenological equations but provide little insight into the molecular basis of the process. Similarly, information on the chemistry of antigenic determinants is obtainable by indirect immunological methods but the way in which the hapten is incorporated into the structure of the membrane remains obscure.

To increase the range of applicable technics, the cell membrane has been isolated and fractionated, but the advances then possible must be weighed against the limitations intrinsic to the processes of isolation and fractionation. A fully satisfactory membrane preparation would retain the full potential biological activity of the membrane and, while lacking none of the native components, would contain no contaminants. This ideal is not easy to attain as the living membrane is not a static structure with a fixed quota of molecules —the molecular constituents range from those whose presence is necessary for gross structural integrity to others that are only very loosely bound and whose association with the complex may be transient yet imperative for the complete fulfillment of membrane metabolism. These latter molecules could well be lost during membrane isolation and, perhaps, be replaced by others having no place in the living membrane. Furthermore, even if a chemically complete membrane is recovered it is usually in fragments, its spatial relationship to the cell is lost, and the most characteristic features of the plasma membrane as the structure controlling the entry and exit of molecules to and from the cell are destroyed. The erythrocyte is an unusually favorable object of study in this respect as conditions are known which allow the contents to be expelled and the membrane to reseal and reform a ghost exhibiting the vectorial attributes of the living cell.

The third stage, the fractionation and solubilization of the isolated membrane, permits the detailed analysis of the membrane as a supramolecular complex. The greater part of this review is concerned with this aspect, but

---

From the Department of Zoology, University of Edinburgh, Edinburgh, Scotland.
Supported by the Medical Research Council.

first we shall consider the various methods that may be applied to the living cell in investigating the contribution proteins can make to its membrane. Attention will be largely confined to proteins of the erythrocyte membrane and even this limited objective cannot be exhaustively treated in one review. A fairly wide field has been surveyed, in places cursorily, to emphasize the vast range of disciplines being brought to bear on the subject. The references chosen are primarily those which will facilitate the amplification of any particular facet. A number of reviews are available that will enable a reader to fit the study of the erythrocyte membrane protein into its place in the general context of membrane structure and functions.[7,10,17,24,31,37,44,57,69,70]

STUDY OF THE INTACT CELL AND INTACT MEMBRANE

Permeability Studies

The information permeability studies provide on the arrangement of protein in the membrane tends to be of two different types. Analysis of active transport systems provides an insight into the detailed arrangement of a small number of protein molecules, i.e., those involved in the transport system, but not on the general arrangement of proteins within the membrane. Study of the effects of proteins on passive ion fluxes, which tend to depend more on the general architecture of the membrane, can give some inkling of the general organization of protein.

Thus, the work of Whittam and others[47,83] has revealed that the ATPase of the sodium/potassium pump is asymmetrically arranged in the membrane, for it is stimulated by external potassium and by internal sodium. This information, however, pertains to a vanishingly small portion of the membrane protein—the pump of human red cells is inhibited by as few as 200 molecules of ouabain.[25]

The nature of the conclusions reached from studies of passive permeability is illustrated by the work of Passow and his associates,[44,45] and Rothstein and his.[58] The implication of proteins stems from the observation that the exceptionally high permeability to chloride and sulfate ions (relative to cations) is pH dependent and diminished at high pH where amino groups titrate. The importance of amino groups is confirmed by the effect of fluorodinitrobezene (FDNB) which at low concentrations causes a pronounced diminution of anion flux. (At much higher concentrations, this reagent also affects cation flux.) The relevant groups in anion transport are near the cell surface for they are blocked by isothiocyanostilbene sulfonate (SITS) which does not enter the cell and, it may be noted, has no effect on cation movements. The effects of FDNB are mimicked by 2-methoxy-5-nitrotropone (MNT), a reagent claimed to be specific for amino groups. Classical sulfhydryl reagents do not affect the flux. There is, therefore, within the membrane one category of superficial amino groups affecting anion flux. Cation flux is influenced by more deeply buried amino groups and by deep seated membrane sulfhydryl residues. The location of these sulfhydryls has been deduced from the time lag required to titrate them with the slowly penetrating p-chloromercuribenzene sulfonate (PCMBS), yet they can not belong to the cell interior as they are not available for interaction with the internal hemo-globin. They differ from the sulfhydryls associated with sugar transport as these latter are superficial and are rapidly labeled by PCMBS. The conclusion that sulfhydryls are present in two locations has been recently confirmed with an electron spin resonance sulfhydryl label which reveals that some molecules of the bound label are weakly immobilized and others strongly immobilized, the latter possibly being in a hydrophobic environment.[59] It is tempting to suggest that the two categories discovered by the two methods are identical, but there is no direct evidence for this possibility.

Spectroscopic Methods

X-ray diffraction of crystalline protein derivatives remains the only un-equivocal method for the complete conformational analysis of proteins. As various spectral characteristics are conformation dependent, much effort has been directed towards squeezing conformational information out of such spectra because the technics are simpler than X-ray diffraction (even if deceptively so) and may be applied to noncrystalline structures. The effects of conformation on spectra have been interpreted by comparing the spectra of proteins of different conformations as determined by X-ray diffraction, the usual model systems being polyamino acids which can be converted from one conformation to another at will. Unfortunately, the extrapolation of these results to globular proteins is not straightforward. The polyamino acids can have simple conformations which may not be readily compared with globular proteins that consist of complex mixtures of short sections of various secondary structures held in tertiary array by stretches of far from random "random coils." The validity of this extrapolation can only be assessed critically in those very few globular proteins with completely analyzed con-formations. Little value can be attached to conclusions based on only one spectroscopic method (see for instance the quite anomalous circular dichroism of immunoglobulin[13]), but they become more meaningful when several technics point in the same direction. In the light of these reservations it will be apparent that application of these methods to membranes, although perhaps technically simple, is extremely hazardous, especially as the spectra may be complicated by light scattering from insoluble membrane particles.

Scattering is least troublesome in the infrared region. Fortunately, the infrared spectra show some conformational dependence, the most readily applicable being the position of the Amide I band which is at 1630 cm^{-1} for β sheets and around 1660 cm.$^{-1}$ for the α helix and random chain. There-fore, spectra of erythrocyte ghosts (Fig. 1) should reveal any extensive β structure. In fact, none can be seen in either dry films or ghost suspensions in heavy water.[39] A similar result is obtained from other plasma mem-branes[33,76] but not from mitochondria.[77] The absence of β sheet disproved a commonly held opinion to the contrary, but the result is rather negative and consequently other spectra, notably optical rotatory dispersion (ORD) and circular dichroism (CD) have been explored chiefly to estimate the amount of α helix present. Using ORD, the α helix content can be estimated from the b_o term of the Moffit-Yang equation from measurements in the visible

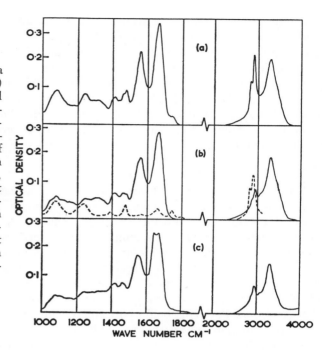

Fig. 1.—Infrared spectra of air-dried specimens. (a) Red-cell ghosts. (b) Solid line, extracted ghost protein; broken line, estimated relative contribution of lipid to spectrum of ghost. (c) Ghost protein denatured with ethanol, showing about 50 per cent of β-conformation. (Reproduced by permission from Science 150:1616-1618, 1965. Copyright 1965 by the American Association for the Advancement of Science.)

region, and a simple unqualified interpretation of the value obtained from ghosts gives a content of 17 per cent. Most recent work has been carried out in the far ultraviolet in the region of the Cotton effects of the peptide bonds. These latter results have been repeatedly described (see for instance Chapt. 1 of reference 10) and the anomalies of the spectra ingeneously interpreted in terms of the arrangement of protein in the membrane. At these short wavelengths, however, scattering effects are, of course, enormous, and it has recently been demonstrated that for both ORD and CD the anomalies are more simply explained by the particulate nature of the material.[28]

Other spectroscopic methods, especially nuclear magnetic resonance and electron spin resonance, hold great promise as technics for membrane study. Reference to their at present limited application to erythrocyte ghost protein is made elsewhere in this review.

Mechanical Properties

The concept of structural proteins in membranes, i.e., proteins whose prime function is a structural contribution to the general architecture of the membrane, has been canvassed for many years. The attribution of this function to an isolated protein is extremely difficult as the manifestation of structural activity depends on the protein being incorporated into some complex. An unequivocal recognition of a structural role for a particular isolated protein is only possible in rare instances, as, e.g., when its addition to some other molecule(s) causes a reorganization to a more native state such as the regeneration of a biological activity other than any catalytic activity the molecule in question might itself possess. Molecules should certainly not be termed structural merely because they lack any apparent catalytic function. The foregoing is not intended to imply any disjunction between structural and catalytic activities. Indeed, proteinaceous structural elements of membranes may all have catalytic activity, serving a dual role, as has been described for cytochrome b in mitochondria. In addition to being a component of the electron transfer chain, this protein, in some undetermined fashion, influences the linkage of succinic dehydrogenase with ubiquinone without itself being either oxidized or reduced at this stage. The discovery of this second role also illustrates the effects of differing solubilization procedures to which we shall refer later. Cytochrome b solubilized by sodium dodecyl sulfate (SDS) retains only the ability to couple succinate dehydrogenase with ubiquinone, while that prepared by guanidine in the presence of glycerol and succinate can also take its place in the electron transfer process.[8,84]

A number of lines of evidence imply that plasma membrane proteins have a structural role in a narrower sense, namely, a mechanical function. There are several reports that the action of proteolytic enzymes on the cell exterior reduces the force required to deform the cell[12,82] and it has been observed that proteins have a protective influence on erythrocytes under osmotic stress.[29] In addition it has been found that 6–10 hours after exsanguination the deformability of red cells falls markedly and the fall is affected by adenosine triphosphate (ATP) and calcium, presumably through a sol-gel transformation involving protein. These latter effects are a property of the membrane as they are detectable in reconstituted ghosts (see accompanying review by Dr. LaCelle). It is noteworthy that the many similarities between black lipid membranes and cell membranes do not extend to their mechanical properties, the natural membranes being in this respect more akin to protein monolayers.[38,51]

Electrophoretic Studies

The essential electrophoretic measurement is the mobility of a cell in an electric field, a property determined by the charged groups in the hydrodynamic slip plane of the cell. Within the limitations of several underlying assumptions,[1,37] the residues responsible for the potential can be identified from this measurement under various conditions and their density within the slip plane calculated. (It must be realized that the conclusions derived from cell electrophoresis are restricted to one specific region of the surface, the hydrodynamic slip plane, or groups effective in that plane, which, at physiological ionic strengths, means groups within 8 Å of the plane.)

An extensive literature exists on the electrophoresis of erythrocytes. Briefly, the identity of the ionogenic groups has been established by studying the effect on mobility of pH and of reagents known to block or remove surface residues. The human erythrocyte has a constant negative charge above pH 4.5 but below this pH the mobility falls as acidic groups on the surface are titrated until at pH 2.0 the mobility is zero. Further decrease in pH does not cause a charge reversal, indicating an absence of cationic groups, a conclusion confirmed by the zero mobility observed after esterification of the surface[64] and the failure of tosylation to affect mobility.[63] The acidic groups are believed to be two types of carboxyls; 61.5 per cent are sialic acid residues removed by neuraminidase, and the remainder are probably protein α carboxyls.

From the study of ion flux, it has been noted earlier that there is on the outer surface of the cell a zone with a preponderance of positive charges, but now it is apparent that the very outermost region is purely anionic. The superficial location of sialic acid residues is confirmed by nuclear magnetic resonance spectra of ghosts.[11]

Electron Microscopy and Immunology

The interpretation of an electronmicrograph in molecular terms is fraught with difficulty and cannot be discussed briefly. It is given special consideration in the review by Weinstein in this issue. The vast body of information on the chemical structure of antigenic determinants on the erythrocyte surface is also unsuitable for cursory examination, but the manner in which the haptens might be built into the membrane is discussed in a later section.

FRACTIONATION OF ERYTHROCYTE MEMBRANE

Study of the intact cell is of paramount importance as it provides the only unequivocal means for observing the natural functioning of the living membrane. Such studies have already permitted the identification of several catalytic sites and the approximate location of various proteinaceous zones within the membrane. The functioning of the whole, however, will not be fully understood until the functioning of the component parts has been elucidated. The process of dissociation of a membrane into these components has been denoted "solubilization." The use of this term is rather unfortunate as it has a strict physical significance that is of secondary importance when what is primarily meant is "dissociatng the membrane into its component parts". The concept of solubility is not easy to apply to macromolecules and in practice frequently means little more than that the particles are not sedimented after, say, three hours at 100,000 × g. Workers are intent on "solubilizing" membranes largely because the vast majority of technics for fractionating any group of particles is preferably or exclusively applicable to molecules in solution. As knowledge of protein behavior in nonaqueous media is limited, and biological activity is usually manifested in aqueous media, this effectively implies converting the membrane into water soluble entities. The aim is to divide the membrane into biologically significant fractions, not simply to break it up into very small pieces that do not sediment, as for instance by prolonged ultrasonication, or merely to disperse it in media of high density such as 8 M urea. There are two major difficulties in the fulfillment of this aim: the complex nature of the forces responsible for membrane stability, and the meaning of the term "biologically significant fractions."

Membrane structure, as opposed to molecular structure, depends upon interactions between molecules, and therefore by definition does not rely on such strong interactions as covalent bonds, but on weak forces. Three sets of interactions must be considered, lipid–lipid, protein–protein and lipid–protein (the carbohydrate components can be treated as conjugated lipids or conjugated proteins). The forces involved are: (1) electrostatic forces that can be either attractive (in which category we may include the hydrogen bond) or repulsive depending on the polar groups concerned, (2) van der Waals forces that arise between nonpolar residues and are highly distance dependent, and (3) hydrophobic forces that are the consequence of interactions between apolar regions and water. (An important result of this last effect is the forcing together of apolar parts of molecules to form nonaqueous micro regions within the bulk aqueous medium as have been demonstrated in aqueous dispersions of lipids and within protein molecules.) It is important to stress that these forces not only hold the membrane together but persist in a modified form in aqueous preparations of the isolated molecules. Thus, any solubilized membrane component will inevitably not be in the conformation in which it existed within the membrane if the isolation procedure has involved any redistribution of the apolar regions as is inevitable when protein has been separated from lipid, and might also occur in other circumstances. The fractionation of any complex held together by a combination of electrostatic and hydrophobic forces is complicated by the fact that mild treatments intended to diminish either force will not impair and might enhance the other.

It can be argued that all constituent molecules of membranes are biologically significant fractions and certainly it will be necessary to have information on the detailed molecular constitution. One could, however, end up with pure preparations of all the constituent molecular species (as is already possible for the lipids) and still not know how they are put together in the membrane originally. Consequently, it might be preferable as a first step to dissociate the membrane not into its separate molecular components but into its biologically functional units, in the hope that these would be easier to study than the complete envelope. Conversely, it might transpire that the manipulation of these supramolecular entities, especially their purification, is extremely difficult and more rapid progress can be achieved by the reconstitution of functional units from their isolated constituent molecules. Successful reconstitution requires much perseverence and good fortune, but no more than does the isolation of a *pure*, biologically active, supramolecular complex.

The problem of fractionation is complicated by the presence in a membrane preparation of loosely bound molecules which may or may not be true components of the functional membrane. These components assume a twofold significance. First, quite apart from any consideration of their role in membrane function, the existence of such molecules renders it imperative that when the chemical composition of an isolated membrane preparation is discussed it is realized that the product depends not only on its original native state, but also on the methods employed in its preparation. Second, there is the question of whether a particular molecule is, or is not, a component of the intact functional membrane. The distinction is in many cases one of degree and comparable with the difference between a prosthetic group and a coenzyme. What is desired at the present juncture is not some arbitary criterion but a careful exploration of the conditions governing the association of a given molecular species with the membrane complex. Of the many enzymes that have been reported to be associated with the ghost, few claims have been subjected to adequate critical investigation. The type of investigation required is exemplified by Mitchell and Hanahan's[42,43] study of the binding of aldolase and glycerol phosphate dehydrogenase to human erythrocyte stroma where it was found that the association was affected by both pH and tonicity. Acetylcholine esterase was much more intimately bound and was not released by any of the conditions that released the other two enzymes. When these workers turned their attention to bovine erythrocyte stroma, they found an entirely different situation. In this species, acetylcholine esterase was readily removed by hypotonic solutions if the stroma was first depleted of calcium. This dissociation was irreversible, in contrast to the trace of hemoglobin remaining with the ghost prepared at pH 6.0 which was reversibly removed at higher pH.[9] The solubilization of acetylcholine esterase is also of interest for it illustrates the possibility of isolating not a protein, but a lipoprotein complex. The isolated particle has a molecular weight of about 2×10^6. Although its activity is 200 times that of the membrane, it is grossly contaminated as most of the lipid is not necessary for activity.

In practice, two attitudes exist towards the problem of solubilization or fractionation, either to dismantle the membrane by sequential operations of increasing severity, or by a one-step solubilization of the whole complex (Table 1).

Methods for Partial Disaggregation of Membrane

These methods are usually distinguishable from those for total solubilization in that they stop short of the application of an organic solvent. Consequently, they tend to separate proteins or lipoproteins from each other, but not to separate proteins from lipids and so they avoid complications ensuing from rearrangement of the native relationship between proteins and lipid. As electrostatic forces are relatively unimportant in lipid–protein interaction, most of these methods depend on altering the balance of electrostatic forces so as to release some membrane component into solution. One cannot assume, however, that an alteration of the electrostatic conditions in the membrane is without effect on the lipid–protein interactions or that removal of one component does not at the same time alter the insoluble residue.

The methods falling into this category mostly depend on the extraction of membranes by media of high or low ionic strength and/or variation of the pH. Thus, ghosts have for some time been known to fragment in distilled water and one of the particles released has been shown by Harris[22,23] to have a complex and characteristic ultrastructure (Fig. 2). The intact entity consists of a hollow cylinder composed of a stack of four rings, each built from 10 protein subunits. The tetramer predominates in preparations from fresh ox blood, but both tetramer and single ring forms are common

Table 1.—Methods for "Solubilization" of Erythrocyte Membrane Proteins

Solvent	Result	Reference
Partial Solubilization:	Releases component of highly characteristic morphology	Harris[22]
Water and freezing		
Chelating agents	Releases fibrous element	Steers and Marchesi[68]
Water after removal of all traces of ions	Membrane dissociates releasing complex mixture of protein—associated with a little lipid	Mazia and Ruby[41]
Hypotonic saline with variation of pH	Various enzymes released from stroma—see text	Mitchell, Mithcell and Hanahan[42] Burger, Fujii and Hanahan[9]
Hypertonic saline	Releases considerable amount of protein + lipid	Mitchell and Hanahan[43] Rosenberg and Guidotti[56]
Warm aqueous phenol	Releases small percentage of total protein, rich in sialic acid and blood group substances.	Springer, Fletcher and Pavlovskis[66]
Dilute acetic acid	Liberates 30–40% of ghost protein as heterogeneous mixture of low molecular weight	Maddy and Kelly[40]
Detergents:	Isolated protein compared with mitochondrial structural protein—for comment see text	Richardson, Hultin and Green[54] Schneiderman and Junga[62]
Cholate and Deoxycholate Triton × 100		
SDS	Requires further investigation in the light of subsequent publications on Mycoplasma and rat liver plasma membranes. Used in conjunction with gel electrophoretic method of molecular weight estimation	Bakerman and Wasemiller[3] Berg[4]
Total Solubilization:	Protein and lipid separation almost complete. Protein water soluble	Maddy[36] Rega et al.[53] Zwaal and van Deenen[88]
Butanol		
2-chloroethanol	Protein and lipid separated by Sephadex LH20 chromatography. Protein precipitates if organic solvent removed	Zahler, Wallach and Luscher[85]
Diethylether	Extract at −20°. Residual protein solubilized by very vigorous methods	Rosenberg and Guidotti[55]
Pentanol	Membrane converted to water soluble lipoproteins	Zwaal and van Deenen[87]
Pyridine	Sialoprotein left in solution after removal of pyridine	Blumenfeld[5]
Urea	Protein fractionated by starch gel electrophoresis	Azen, Orr and Smithies[2]

Table 1.—Methods for "Solubilization" of Erythrocyte Membrane Proteins (Cont'd)

Solvent	Result	Reference
Urea + butanol in acid	Solubilized protein may be separated into an antigen rich sialoprotein fraction and a fraction poor in sialic acid and lacking the blood group antigens	Poulik and Lauf[49]

in preparations from human cells. This structure has now been observed in human, ox, sheep, guinea pig, dog, chicken and turkey erythrocytes. Its biological function is not yet known. The effects of water may be extended by the addition of a chelating agent to remove divalent cations from the membrane and so destroy any structures stabilized by divalent ion bridges. It has been reported that such treatment releases one protein–*spectrin* representing up to 20 per cent of the total protein from guinea pig erythrocyte membranes. This protein undergoes polymerization in 0.1 M potassium chloride into fibers reminiscent of actin fibers.[68] A system of this type could conceivably be involved in the mechanical changes controlled by ATP. If still greater precautions are taken to remove all ions,[41] the whole membrane complex disintegrates at pH 9.0, most of the protein, associated with a little lipid, passing into solution, and the bulk of the lipid forming a water insoluble

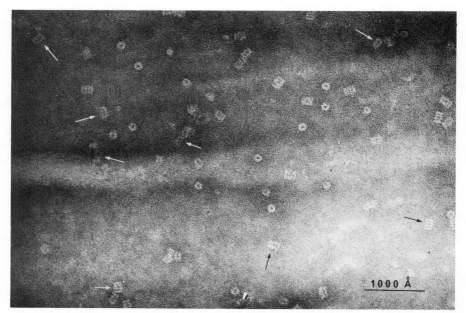

Fig. 2.—A preparation of the tetramer (hollow cylinder) protein purified by sucrose density gradient centrifugation from human erythrocyte ghosts. Negative stain, 2 per cent uranyl acetate. Scale marker = 100 nm. (Courtesy of Dr. J. H. Harris).

phase. Mazia and Ruby[41] note certain affinities between this protein and similar ones isolated from other cell structures, e.g., microtubules, and suggest that they all belong to a special class of structural proteins–*tectins*. Rather more information is required before the validity of this interpretation can be assessed.

The opposite type of treatment, i.e., use of high ionic strength solutions, also liberates molecules from the membrane. Particular care must be exercised in the interpretation of these data because such solutions would release any electrostatically bound cytoplasmic contaminants, as clearly happens with liver plasma membrane preparations.[15] Prolonged extraction with sodium chloride between 1.0 and 2.0 M releases much lipid and protein from human ghosts,[43,56] including most of the acetylcholine esterase. In view of the previously quoted work, the release of this enzyme implies a fairly extensive disruption of membrane structure by strong salt.

Recently Maddy and Kelly[40] have employed 0.26 M acetic acid as a selective solubilization agent for bovine ghosts. It is selective in that it does not liberate any sialic acid or lipid from ghosts as long as they have not been previously washed with water. This method differs from those listed above as the conditions include low ionic strength below the isoelectric point of the membrane proteins and the proteins are liberated as cations. A complex mixture of proteins has been obtained in the four species examined (ox, human, sheep, guinea pig) each exhibiting a complex pattern of bands on acrylamide gel electrophoresis. In the ox, the species most widely studied, about 30 per cent of the ghost protein is solubilized (Fig. 3).

Total Solubilization

In the preceding section, methods which dissociated various entities from the membrane and left an insoluble residue were described. In this section, methods which solubilize the whole complex leaving no insoluble residue are considered. These methods can, of course, be applied to any of the residues resulting from partial solubilization, as has been recently done by Rosenberg and Guidotti[56] who solubilized the final proteinaceous residue resulting from extraction of ghosts by chelating agents, followed by high salt extraction and lipid removal with sodium dodecyl sulfate (SDS). All methods in this category involve separation, or at least rearrangement of lipid and protein, a process that necessarily modifies the conformation of the proteins concerned, and the relative merits of the different methods depend on the extent of any irreversible damage done.

Organic Solvents. When the organic solvents that have been used to solubilize the membrane are surveyed, it is not possible to discern any common factor responsible for the solvent effect, except possibly the ability to dissolve lipid. Thus, solvents can be either totally or partially miscible with water, the former usually being well known non-aqueous protein solvents. For example, 2-chloroethanol, as it is a common solvent for lipids and proteins, dissolves membranes, and as the components remain dissociated in the solution, the lipid can be separated from the protein by passage through a Sephadex LH20 column.[85] Butanol, on the other hand, is only partially mis-

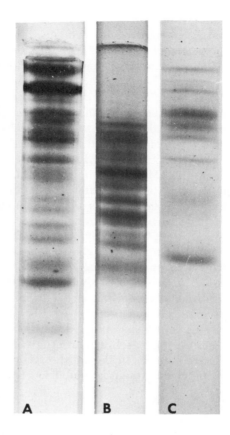

Fig. 3.—Acrylamide gel electrophoresis of acetic acid soluble proteins of erythrocyte membrane. (A) ox, (B) human, (C) guinea pig.

cible with water, and when a membrane suspension is shaken with an excess of butanol the protein remains in the aqueous phase while virtually all the lipid passes into the organic phase.[36] Most surprisingly, the chemically similar pentanol has a quite different effect on membranes. Instead of producing water soluble protein free of lipid, the result is a mixture of water soluble lipoprotein(s).[87] As the water miscibility of the solvent can vary, so also can its polarizability. Many solvents, as was pointed out by Zahler et al[85] are acidic or are used under acidic conditions, but butanol and pentanol are used near neutrality and the basic pyridine can also, as a 30 per cent aqueous solution, dissolve red cell ghosts. Removal of the pyridine causes precipitation of a lipid–protein complex but leaves the sialic acid bearing proteins in solution.[5]

Detergents. Detergents have been successfully employed as solubilizing reagents, but they are also powerful denaturants at low concentrations.[73] The literature indicates that they must be employed with extreme circumspection. Such caution is largely actuated by the extensive and still unresolved investigation of "the structural protein" of mitochondria. It will be remembered that, using a complex mixture of detergents, the isolation of a "pure" protein with no catalytic activity from mitochondria and membrane systems including the erythrocyte was reported.[54] The protein was present in significant amounts, lacked catalytic activity, was prone to aggregate, was rich in hydrophobic amino acids and on these grounds was designated a "structural" protein. It should be noted that in addition to the detergent treatment, isolation of the structural protein usually involved washing with a warm organic solvent to remove the detergent. Many of the early claims made for the preparation have been modified[18] and, most significantly, the preparation has been shown to contain at least one denatured, inactivated catalytic protein—the F_1 ATPase[60] and partially inactivated F4 protein.[50] These investigations of the structural protein are of twofold significance for they illustrate that detergents can inactivate proteins and also can cause them to form highly stable intractable aggregates. This salutary tale must, however, be weighed against the happier results emanating largely from Racker's laboratory on the isolation of cytochrome b (see above) and other components of the respiratory chain.

In addition to their use in isolating proteins, detergents have been employed as agents for the dissociation of membranes into soluble lipoproteins which, it has been repeatedly claimed, represent the lipoprotein subunits of the membrane. In the case of the plasma membrane, sodium dodecyl sulfate has been used for *Mycoplasma*,[16] liver cells,[6] and red blood cells.[3] In the first two cases, the conclusion that the homogeneous peak observed in the ultracentrifuge was a "membrane lipoprotein subunit" was shown to be nothing more than a fortuitous coincidence of lipid and protein. When the "subunit" was analyzed on a density gradient, the lipids and proteins separated. On removal of detergents, the proteins and lipids reassociated in the presence of magnesium. The most recent evidence[52] has shown this process to be a complex multistep phenomenon. The effects of SDS on the erythrocyte membrane have not been subjected to the critical examination afforded to *Mycoplasma,* and perhaps more relevantly the liver membrane, and until this has been done any claims must be treated with scepticism.

Acrylamide gel electrophoresis of proteins in SDS is now advocated as a method for determining their molecular weights. The technic depends on the linear logarithmic relationship discovered between the molecular weights of polypeptides of known molecular weight and their relative mobilities in gels.[65,80] Not surprisingly, the method has been applied to membrane proteins.[4,30,61] Different membranes dissolved in SDS and electrophoresed in its presence undoubtedly give complex and characteristic band patterns, but it is not unequivocally established that each band represents a species of protein of the molecular weight expected by comparison with the polypeptides of known weight. The molecular weight of these polypeptides is known because they can be readily brought into solution which in turn is a consequence of their not undergoing complex interactions with other molecules. The very existence of such interactions is the major difficulty in handling membrane proteins. It is not *necessary* that SDS completely destroys them and fully dissociates the proteins. The further exploitation of this technic could, however, be of considerable interest.

Biological Significance of Isolated Membrane Fractions

In the first part of this review, the range of functions that might be expected of ghost proteins was considered, in the second the isolation of proteins from the ghost was discussed. We shall now attempt to assess to what extent the isolated proteins may be assigned to the functional characteristics of the intact membrane and to evaluate the status of the protein fractions obtained by the various methods. It will be appreciated that many of the problems in this correlation arise from the disappearance of the vectorial components in a protein's behavior once it is isolated, and the difficulty of identifying the original location of an isolated protein in the intact membrane. The former difficulty might be mitigated by the discovery of other reagents analogous to ouabain that have specific effects on a vectorial process in a cell and on an apparently scalar process in the isolated membrane. The second problem might be solved by deducing the original location of proteins by tracing markers such as sialic acid, all of which is on the outer surface of the erythrocyte, or introducing a tag like SITS to only one side of the intact membrane.[4,35] It will be only too apparent from the ensuing paragraphs that the chasm between work in the intact cells and on isolated fractions is wide and deep, and only in a very few instances can an isolated fraction be related back to its role and position within the ghost.

Enzymic Activities

Pennell[46] presented a comprehensive list of the enzymes that have been detected in erythrocytes and considered the information then available on their distribution between membrane and cell contents. Later work dealing critically with this latter problem has already been discussed in an earlier section. In the course of an ideal system of membrane fractionation, the full enzymic content of the complex should be recovered in its fractions. Failure to attain this ideal can be attributed to two major causes. In the first place, many of the fractionation schemes are destructive and loss of activity is due to denaturation of the enzyme protein. Indeed, retention of enzymic activity is perhaps the most rigorous criterion for evaluating any disruptive procedure. Unfortunately, the second factor, namely the requirement of a lipid component for activity, precludes any simple assessment, for activity can be lost, not as a consequence of any irreversible deleterious effect of the solubilization procedure on the protein, but merely because some requisite lipid component has been removed. The lipoprotein complex exhibiting acetylcholine esterase activity has already been mentioned, membrane NADH reductase is another,[86] but by far the most widely explored example is the sodium/potassium dependent ATPase.[26,47,83] In this case, many workers have claimed to have reactivated the enzyme by incubation of the apoprotein with various lipids, and the disparities between the reports illustrate the difficulty of the operation. Different workers find different lipids to be effective, although sometimes the regeneration is only partial. Even if full activity is restored, it remains to be shown that the original native state, rather than an active alternative, has been reconstructed. Recently, the factors responsible for the lipid reactivation of the enzyme prepared from brain have been systematically analyzed.[72] The essential features are a negative charge, preferably a phosphate but a carboxyl or a sulfate are also effective, and two fatty acyl chains, optimally 10 carbon atoms long.

It is to be hoped that in the near future the enzymic activities of the many membrane proteins that have now been isolated will be fully explored.

Antigenic Activity

The immunological studies of the intact cell to which brief reference was made earlier have elucidated much of the detailed chemistry of the actual haptens but have shed little light on the way in which the haptens are built into the structure of the membrane. The advances that can result from fractionating the membrane may best be illustrated by the ABO system of human cells. Early work on these groups in red cells reported activity in organic solvent extracts, and it was concluded that the active materials were lipoidal. With the discovery of soluble forms of the antigens in various mucopolysaccharide secretions, erythrocytes fell into disfavor as a source, although it was tacitly assumed that the molecules in the membrane would be the same as those in the secretions—mostly carbohydrate with a small amount of peptide.[78] When interest in the red cell was renewed, the antigenic activity was, surprisingly, reported to be in the glycolipids,[21] not the proteins. Still more recently it has been realized that the activity recovered in these glycolipids was too low to account for the amount present in the cell, and when Whittemore et al[79] tested butanol solubilized protein after careful re-extraction to remove the final traces of lipid, they found positive reactions for AB and H groups. Therefore, the AB haptens can be attached to both lipids and proteins. The MN system is also associated with butanol solubilized protein.[48] The MN positive extracts prepared by a somewhat drastic phenol method contain much protein[65] and an active glycoprotein may be released from the cells enzymically.[74] Rhesus antigens represent another arrangement, as activity depends on the integrity of a lipid–protein complex.[19] While butanol extraction of lyophilized ghosts destroys antigenic activity, it can be restored by readdition of the lipid extract to the protein residue. The antigenic specificity resides in the protein, the lipid activation being relatively nonspecific with respect to the lipid.[20]

Antigenic activity associated with red cell ghost protein has been detected in other species. As with man, the blood group antigens of cattle fall into two categories, those present in the protein after butanol fractionation and those destroyed by this treatment. As these latter cannot be detected in any of the fractions, their expression may depend on the integriy of some lipoprotein complex, although we are not at present able to regenerate activity by incubation of the protein with the lipid.[67] Two similar categories exist in the sheep as exemplified by the D antigen which retains its activity after both lyophilization and butanol treatment (it is found with the protein) and the M antigen which is labile and seems to require the integrity of the membrane or some supramolecular component.[32] The study of this antigen is of major relevance to the organization of protein within the membrane as M antigen is found in all HK (high potassium) erythrocytes and only some LK (low potassium) cells. Another antigen, the L antigen, is found exclusively in LK cells, and antiserum to this antigen stimulates the ion pump in LK cells.[14] Treatment with anti-M serum does not affect the pump. The molecular basis of the difference between the two cells is not understood. No difference is detectable between the lipids, and none apparent between the isolated proteins either, but this finding probably illustrates the inadequacies of the methods available for fractionating the proteins rather than anything else. HK sheep have more ouabain binding sites than do LK, but a greater number of Na^+/K^+ pump sites does not fully explain the difference between the two cells as other kinetic differences have also been observed.[75]

In addition to studying blood group antigens, immunologists can make another contribution to membrane study by using their technics to test the homogeneity of protein fractions derived from membranes.[27,48]

Spectroscopic Methods

Although a molecular interpretation of membrane spectra is ambiguous, spectroscopy can be used to detect changes in membrane derivatives caused by solubilization even if the molecular significance of the alteration is not readily apparent. Protein in the intact red cell ghost has no detectable β structure in the infrared, neither does its protein after solubilization by butanol.[39] Protein solubilized by 2-chloroethanol or dimethylformamide, however, shows a β Amide I band.[11] A different application is illustrated by the effect of electron microscope fixatives on the ORD and CD spectra of membranes which is so drastic as to throw doubt on any detailed molecular interpretation of micrographs obtained through their use.[34] As the understanding of the spectra of complex intermolecular associations improves, spectroscopic studies can be expected to be of increasing value in membrane study.

Attempted Comparison of Results Obtained by Different Methods

The extent of our ignorance of the proteins that go into the making of the erythrocyte membrane is glaringly exposed when an attempt is made to compare the results obtained by the various procedures in current use on different species. The only firm conclusion one can draw is that the proteins are capable of highly complex interactions with each other. There is not even general agreement on the number of proteins involved, for the number of apparent fractions is largely a consequence of the method used to isolate the proteins. Thus, protein solubilized by 2-chloroethanol precipitates if the organic reagent is dialyzed away into water. Although the protein has a low sedimentation coefficient in the chloroethanol, it can apparently only be fractionated under somewhat drastic conditions.[71] By contrast, protein solubilized by butanol remains in solution after removal of the residual butanol, yet this apparently innocuous solution is remarkably resistant to fractionation. Zwaal and van Deenen[88] have succeeded in fractionating it electrophoretically under acidic conditions in the presence of urea but in our hands, under these same conditions and even under Takayama's conditions, although several bands may be detected within the gel, most of the protein remains at the origin.

Our recent solubilization with acetic acid of ox ghosts has the advantage that the proteins are readily fractionated by acrylamide gel electrophoresis using acetic acid as the electrophoresis solvent—a relatively mild procedure. In the analytical centrifuge, the protein solution in acetic acid shows two partially resolved components at 0.7 and 1.7S (1 per cent protein concentration), but if 0.1 M potassium chloride is added the proteins aggregate reversibly to 4S and 13S. Although these proteins must be present in the protein solubilized by butanol, the butanol method yields an aqueous solution of 10 and 20S particles. We have shown that under these conditions the acetic acid soluble proteins (M_A) are associated with other proteins (M_R) to form the 20S particles. The M_R proteins we can isolate only as an insoluble residue. The 10S component can be isolated as a pure fraction and as it contains most of the carbohydrate of the membrane, especially the sialic acid, it probably forms the outermost layer of protein in the intact membrane. It seems not unlikely that this scheme has certain affinities with that based on a butanol extraction in the presence of formic acid.[49] When ox and human ghosts are extracted with ATP and β-mercaptoethanol, i.e., the conditions recommended for the isolation of spectrin (about 20 per cent of the ghost protein) the solution does indeed contain a component that aggregates as fibrils on incubation in higher concentrations of ATP plus magnesium ions. In the electron microscope, however, many small particles are seen that do not aggregate, including the tetramer particle of Harris. This heterogeneity is confirmed by acetic acid electrophoresis which reveals many bands in a pattern similar to that of acetic acid extracts.

The reader will undoubtedly have noted that there are very few bridges crossing the gap between investigations of the intact membrane as described in the first section and the studies reported in the later sections. From this dichotomy, which the writer trusts is primarily a consequence of the present state of the subject rather than of his own ignorance, one hopes the direction of future efforts may be discerned. Students of the intact cell must strive to interpret their results in molecular terms, and those who isolate fractions should, and here the need is pressing, attempt to relate their fractions to the known biological activities of the intact membrane and where possible its structure. Space has precluded any comparison of our understanding of the relationship between structure and function of the erythrocyte membrane with that of other membrane systems. The ghost certainly has many merits as a subject for study, but it must be recognized that its prime function is to act as a container for hemoglobin, and if this be so, a great part of the membrane may appear inert, and consequently a relatively intractable subject.

ACKNOWLEDGMENTS

I am grateful to Dr. J. Harris and Mr. P. Agutter for their helpful comments on the manuscript.

REFERENCES

1. Ambrose, E. J. (Ed.): Cell Electrophoresis. London, Churchill, 1965.

2. Azen, E. A., Orr, S., and Smithies, O.: Starch gel electrophoresis of erythrocyte stroma. J. Lab. Clin. Med. 56:440, 1965.

3. Bakerman, S., and Wasemiller, G.: Studies on structural units of human erythrocyte membrane. I. Separation, isolation and partial characterization. Biochemistry 6:1100, 1967.

4. Berg, H. C.: Sulfanilic acid diazonium salt—a label for the outside of the human erythrocyte membrane. Biochim. Biophys. Acta 183:65, 1969.

5. Blumenfeld, O.: The proteins of the erythrocyte membrane obtained by solubilization with aqueous pyridine solution. Biochim. Biophys. Res. Commun. 30:200, 1968.

6. Bont, W. S., Emmelot, P., and Vaz Diaz, H.: Studies on plasma membranes VIII. The effect of sodium deoxycholate and dodecyl sulfate on isolated rat-liver plasma membranes. Biochim. Biophys. Acta 173:389, 1969.

7. Branton, D.: Membrane structure. Ann. Rev. Plant Physiol. 20:209, 1969.

8. Bruni, A., and Racker, E.: Resolution and reconstitution of the mitochondrial electron transport system. I. J. Biol. Chem. 243:962, 1968.

9. Burger, S. P., Fujii, T., and Hanahan, D. J.: Stability of the bovine erythrocyte membrane. Release of enzymes and lipid components. Biochemistry 7:3683, 1968.

10. Chapman, D. (Ed.): Biological Membranes. New York, Academic Press Inc., 1968.

11. —, Kamat, V. B., and Levene, R. J.: Infra red spectra and the chain organization of erythrocyte membranes. Science 160:314, 1968.

12. Day, T. M., and Maddy, A. H.: The turnover of the fibroblast surface. Exp. Cell. Res. 53:665, 1968.

13. Dorrington, K. J., Zarlengo, M. H., and Tanford, C.: Conformational change and complementarity in the combination of H and L chains of immunoglobulin G. Proc. Nat. Acad. Sci. U.S.A. 58:996, 1967.

14. Ellory, J. C. and Tucker, E. M.: Stimulation of the potassium transport system in low potassium type sheep red cells by a specific antigen-antibody reaction. Nature 222:477, 1969.

15. Emmelot, P., and Benedetti, E. L.: Structure and function of isolated plasma membranes from liver. In: Peeters, H., (Ed.): Protides of the Biological Fluids. Amsterdam, Elsevier, 1967, 15:315.

16. Engelman, D. M.: Solubilisation and aggregation properties of membrane components from Mycoplasma laidlawii. In: Bolis, L. and Pethica, B. A. (Eds.): Membrane Models and the Formation of Biological Membranes. Amsterdam, North-Holland, 1968, p. 203.

17. Finean, J. B.: The molecular organization of cell membranes. Prog. Biophys. Molec. Biol. 16:143, 1966.

18. Green, D. E., Haard, N. F., Lenaz, G., and Silman, H. I. On the non-catalytic proteins of membrane systems. Proc. Nat. Acad. Sci. U.S.A. 60:277, 1968.

19. Green, F. A.: Rh antigenicity: an essential component soluble in butanol. Nature 219:86, 1968.

20. —: Phospholipid requirement for Rh antigenic activity. J. Biol. Chem. 243:5519 1968.

21. Hakamori, S., and Strycharz, C. D.: Investigations on cellular blood group substances. I. Biochemistry 7:1279, 1968.

22. Harris, J. H.: The isolation and purification of a macromolecular protein component of the human erythrocyte ghost. Biochim. Biophys. Acta 188:31, 1969.

23. —: Some negative contrast staining features of a protein from erythrocyte ghosts. J. Mol. Biol. 46:329, 1969.

24. Henn, F. and Thompson, T. E.: Synthetic lipid bilayer membranes. Ann. Rev. Biochem. 38:241, 1969.

25. Hoffman, J. F.: The interaction between tritiated ouabain and the Na-K pump in red blood cells. J. Gen. Physiol. 54:343s, 1969.

26. Hokin, L. E.: On the molecular characterization of the sodium-potassium transport adenosine triphosphatase. J. Gen. Physiol. 54:327s, 1969.

27. Howe, C., and Lee, L. T.: Immunochemical study of hemoglobin-free human erythrocyte membranes. J. Immunol. 102: 573, 1969.

28. Ji, J. H., and Urry, D. W.: Correlation of light scattering and absorption flattening effects with distortions in the circular dichroism patterns of mitochondrial

membrane fragments. Biochim. Biophys. Res. Commun. 34:404, 1969.

29. Katchalsky, A., Kedem, O., Klibansky, C. and de Vries, A.: Rheological considerations of the haemolysing red blood cell. In: Copley, A. L., and Stainsby, G. (Eds.): Flow Properties of Blood and Other Biological Systems. New York, Pergamon, 1960, p. 171.

30. Kiehn, E. D., and Holland, J. T.: Multiple protein components of mammalian cell membranes. Proc. Nat. Acad. Sci. U.S.A. 61:1370, 1968.

31. Korn, E. D.: Cell membranes: structure and synthesis. Ann. Rev. Biochem. 38:263, 1969.

32. Lauf, P. K., and Tosteson, D. C.: The M-Antigen in HK and LK sheep red cell membranes. J. Membrane Biol. 1:177, 1969.

33. Lenard, J., and Singer, S. J.: Protein conformation in cell membrane preparations as studied by optical rotatory dispersion and circular dichroism. Proc. Nat. Acad. Sci. U.S.A. 56:1828, 1966.

34. —., and —.: Alteration of the conformation of proteins in red blood cell membranes and in solution by fixatives used in electron microscopy. J. Cell. Biol. 37:117, 1968.

35. Maddy, A. H.: A fluorescent label for the outer components of the plasma membrane. Biochim. Biophys. Acta 88:390, 1964.

36. —: Properties of the protein of the plasma membrane of ox erythrocytes. Biochim. Biophys. Acta 117:193, 1966.

37. —: The chemical organization of the plasma membrane of animal cells. Int. Rev. Cytol. 20:1, 1966.

38. —: The organisation of protein in the plasma membrane. Symposium Int. Soc. Cell Biol. 6:255, 1968.

39. —, and Malcolm, B. R.: Protein conformations in the plasma membrane. Science 150:1616, 1965.

40. — and Kelly, P. G.: F.E.B.S. Letters (in press).

41. Mazia, D., and Ruby, A.: Dissolution of erythrocyte membranes in water and comparison of the membrane protein with other structural proteins. Proc. Nat. Acad. Sci. USA 61:1005, 1968.

42. Mitchell, C. D., Mitchell, W. B., and Hanahan, D. J.: Enzyme and hemoglobin retention in human erythrocyte stroma. Biochim. Biophys. Acta 104:348, 1965.

43. — and Hanahan, D. J.: Solubilization of certain proteins of the human erythrocyte stroma. Biochemistry 5:51, 1966.

44. Passow, H.: Passive ion permeability of the erythrocyte membrane. Prog. Biophys. 19:423, 1969.

45. —, and Schnell, K. F.: Chemical modifiers of passive ion permeability of the erythrocyte membrane. Experientia 25:460, 1969.

46. Pennell, R. B.: Composition of normal human red cells. In: Bishop, C. and Surgenor, D. M. (Eds.): The Red Blood Cell. New York, Academic Press, 1964, p. 29.

47. Post, R. L., Kume, S., Tobin, T., Orcutt, B., and Sen, A. K.: Flexibility of an active centre in sodium-potassium adenosine triphosphatase. J. Gen. Physiol. 54:306s, 1969.

48. Poulik, M. D., and Bron, C.: Immunology of red cell membrane proteins. In: Jamieson, G. A., and Greenwalt, T. J. (Eds.): The Red Cell Membrane, American Red Cross Symposium. Philadelphia, Lippincott, 1970.

49. — and Lauf, P. K.: Some physicochemical and serological properties of isolated protein components of red cell membranes. Clin. Exp. Immunol. 4:165, 1969.

50. Racker, E. and Bruni, A.: The role of phospholipids in structure and function of the mitochondrial membrane. In: Bolis, L., and Pethica, B. A. (Eds.): Membrane Models and the Formation of Biological Membranes. Amsterdam, North-Holland, 1968, p. 138.

51. Rand, R. P.: The structure of a model membrane in relation to the viscoelastic properties of the red cell membrane. In: New York Heart Association Symposium. Biological Interfaces; Flows and Exchanges. Boston, Little Brown, 1968, p. 173.

52. Razin, S., Ne'eman, Z., and Ohad, I.: Selective reaggregation of solubilised mycoplasma-membrane proteins and kinetics of membrane reformation. Biochim. Biophys. Acta 193:277, 1969.

53. Rega, A. F., Weed, R. I., Reed, C. F., Berg, G. G. and Rothstein, A.: Changes in the properties of human erythrocyte membrane protein after solubilisation by butanol extraction. Biochim. Biophys. Acta 147:297, 1967.

54. Richardson, S. H., Hultin, H. O., and

Green, D. E.: Structural proteins of membrane systems. Proc. Nat. Acad. Sci. USA 50:821, 1963.

55. Rosenberg, S. A., and Guidotti, G.: The protein of human erythrocyte membranes. J. Biol. Chem. 243:1985, 1968.

56. —, and —: Fractionation of the protein components of human erythrocyte membranes. J. Biol. Chem. 244:5118, 1969.

57. Rothfield, L., and Finkelstein, A.: Membrane biochemistry. Ann. Rev. Biochem. 37:463, 1969.

58. Rothstein, A.: Membrane permeability of erythrocytes. In: Deutsch, E., Gerlach, E., and Moser, K. (Eds.): Metabolism and Membrane Permeability of Erythrocytes and Thrombocytes. Stuttgart, Georg Thieme, 1968, p. 407.

59. Sandberg, H. E., Bryant, R. G., and Piette, L. H.: Studies on the location of sulphydryl groups in erythrocyte membranes and magnetic resonance spin probes. Arch. Biochim. Biophys. 133:144, 1969.

60. Schatz, G., and Saltzgaber, J.: Identification of denatured mitochondrial ATPase in "structural protein" from beef heart mitochondria. Biochim. Biophys. Acta 180:186, 1969.

61. Schnaitman, C. A.: Comparison of rat liver mitochondrial and microsomal membrane proteins. Proc. Nat. Acad. Sci. USA 63:412, 1969.

62. Schneiderman, L. J., and Junga, I. G.: Isolation and partial characterization of structural protein derived from human red cell membrane. Biochemistry 7:2281, 1968.

63. Seaman, G. V. F., and Heard, D. H.: The surface of the washed human erythrocyte as a polyanion. J. Gen. Physiol. 44:251, 1961.

64. —, and Haydon, D. A.: Electrokinetic studies on the ultrastructure of the human erythrocyte. Arch. Biochim. Biophys. 122:126, 1967.

65. Shapiro, A. V., Vinuela, E., and Maizel, J. V.: Molecular weight estimation of polypeptide chains by electrophoresis in SDS-polyacrylamide gels. Biochem. Biophys. Res. Commun. 28:815, 1967.

66. Springer, G. F., Fletcher, M. A., and Pavlovskis, O.: Homogeneous erythrocyte glycoproteins with blood-group-virus- and endotoxin-receptor activities. In: Peeters, H. (Ed.): Protides of the Biological Fluids Vol. 15. Amsterdam, Elsevier, 1967, p. 109.

67. Spooner, R. L., and Maddy, A. H.: The detection of red cell antigens in ox red cell membrane protein. Proc. Eleventh Europ. Cong. Animal Blood Groups and Biochemical Polymorphism. Warsaw, 1968, Dr. W. Junk, N.V., The Hague, 1970.

68. Steers, E., and Marchesi, V. T.: Studies on a protein component of guinea pig erythrocyte membranes. J. Gen. Physiol. 54:65s, 1969.

69. Stein, W. D.: The Movement of Molecules Across Cell Membranes. New York, Academic Press, 1967.

70. Stoekenius, W., and Engelman, D. M.: Current models for the structure of biological membranes. J. Cell Biol. 42:613, 1969.

71. Takayama, K., Lennan, D. H., Tzagoloff, A., and Stoner, C. D.: Studies on the electron transfer system. LXVII. Arch. Biochem. Biophys. 114:223, 1963.

72. Tanaka, R., and Sakamoto, T.: Molecular structure of phospholipid essential to activate (Na^+ -K^+ -Mg^{2+}) dependent ATPase and (K^+ -Mg^{2+}) dependent phosphatase of bovine cerebral cortex. Biochim. Biophys. Acta 193:384 1969.

73. Tanford, C.: Protein denaturation. Adv. Protein Chem. 23:122, 1968.

74. Thomas, D. B., and Winzler, R. J.: Structural studies on human erythrocyte glycoproteins. J. Biol. Chem. 244:5943, 1969.

75. Tosteson, D. C.: Sodium and potassium transport across the red cell membrane. In: Jamieson, G. A., and Greenwalt, T. J. (Eds.): Red Cell Membrane Structure and Function. Philadelphia, J. B. Lippincott, 1970, p. 291.

76. Wallach, D. F. H., and Zahler, P. H.: Protein conformations in cellular membranes. Proc. Nat. Acad. Sci. U.S.A. 56:1552, 1966.

77. —, Graham, J. M., and Fernbach, B. R.: β conformation in mitochondrial membranes. Arch. Biochem. Biophys. 131:322, 1969.

78. Watkins, W. M.: Blood-group substances: their nature and genetics. In: Bishop, C., and Surgenor, D. M. (Eds.): The Red Blood Cell. New York, Academic Press, 1964, p. 359.

79. Whittemore, N. B., Trabold, N. C., Reed, C. F., and Weed, R. I.: Solubilised glycoprotein from human erythrocyte membranes possessing blood group A, B and H activity. Vox Sang. 17:289, 1969.

80. Weber, K. and Osborn, M.: The re-

liability of molecular weight determinations by dodecyl sulfate-polyacrylamide gel electrophoresis. J. Biol. Chem. 244:4406, 1969.

81. Weed, R. I., LaCelle, P. L., and Merrill, E. W.: Metabolic dependence of red cell deformability. J. Clin. Invest. 48:795, 1969.

82. Weiss, L.: Studies on cell deformability II. Effects of some proteolytic enzymes. J. Cell. Biol. 30:39, 1966.

83. Whittam, R.: The asymmetrical stimulation of a membrane ATPase in relation to active cation transport. Biochem. J. 84:110, 1962.

84. Yamashita, S., and Racker, E.: Reconstitution of the mitochondrial oxidative chain from individual components. J. Biol. Chem. 243:2446, 1968.

85. Zahler, P. H., Wallach, D. H. F., and Luscher, E. P.: Complete solubilisation of plasma membranes and isolation of lipid-free membrane proteins. In Peeters, H. (Ed.): Protides of the Biological Fluids, Vol. 15. Amsterdam, Elsevier, 1967, p. 69.

86. Zamudio, I., Cellino, M., and Canessa-Fischer, M.: A NADH oxidizing system of the cell membrane of human erythrocytes. In: Tosteson, D. C. (Ed.): The Molecular Basis of Membrane Function. Englewood Cliffs, N. J., Prentice-Hall, 1969, p. 545.

87. Zwaal, R. F. A., and van Deenen, L. L. M.: The solubilisation of human erythrocyte membranes by n-pentanol. Biochim. Biophys. Acta 150:323, 1968.

88. —, and —: Protein patterns of red cell membranes from different mammalian species. Biochim. Biophys. Acta 163:44, 1968.

Lipids of Human Red Cell Membrane: Normal Composition and Variability in Disease

By Richard A. Cooper

T HE LIPIDS of mature red cells are confined to the membrane. This review will discuss the composition and steady-state turnover of lipids in normal red cells and deviations from normal under physiologic and pathologic conditions. Emphasis will be placed on human red cells, although, particularly where they appear to provide insight into human disease, the characteristics of animal red cells will also be explored.

COMPOSITION

The major lipids of the human red cell membrane are cholesterol and the phospholipids (Table 1). On a weight basis phospholipids account for approximately 70 per cent of human red cell lipids. The cholesterol:phospholipid mole ratio is approximately 0.8. Glycosphingolipids account for 2 per cent of human red cell lipids by weight,[170] and small amounts of free fatty acids are also present.[158,186] Contrary to early reports,[40] the red cell membrane does not appear to contain either glycerides[170] or sterol esters.[21,170] Analysis of red cell ghosts confirms that all the lipid present in the mature red cell resides in the membrane.[35,40]

The study of red cell membrane lipids requires extraction technics capable of liberating completely stromal lipids into the solvent phase. Low values in early reports have been the result of incomplete extraction technics.[40,57,130] The procedure of Folch and coworkers,[48] commonly used for the extraction of tissue lipids, has proved unsatisfactory because of the large amount of hemoglobin also extracted,[142,146] and it has been modified for use with red cells.[142,186] Certain advantages exist in using isopropanol rather than methanol, since hemoglobin is less soluble in longer chain alcohols.[146] Antioxidants have been employed to prevent the autoxidation of fatty acids during extraction.[37]

Cholesterol

Cholesterol normally accounts for more than 99 per cent of the neutral lipid present in mature red cells[112] and it exists in the free, or unesterified form. Small amounts of plant sterols have been found recently in the erythrocyte membranes of patients receiving diets rich in these sterols.[150a] While the greatest likelihood is that membrane cholesterol is distributed evenly

From Thorndike Memorial Laboratory and Harvard Medical Unit, Boston City Hospital, Boston, Mass.

Supported by Grants HE/AM 07652, AM-05391, FR-76, and 1-F3-AM38,345 from the National Institutes of Health.

RICHARD A. COOPER, M.D.: Thorndike Memorial Laboratory and Harvard Medical Unit, Boston City Hospital, Boston, Mass.

Table 1.–Lipids of the Normal Human Red Cell Membrane

	μg./10^8 cells	μmole/10^{11} cells
Cholesterol*	12.67	32.8
Lipid phosphorus*	1.23	
Phospholipid*	30.72	} 3.97
Glycolipid†	1.12	1.0
Free fatty acid‡	0.81	2.6
		Per Cent of Total Phospholipid
Sphingomyelin		25.2
Lecithin		31.0
Phosphatidyl serine (+ phosphatidyl inositol)		13.5
Phosphatidyl Ethanolamine		27.3
Lysolecithin		1.3
Other (polyglycerol phosphatide, phosphatidic acid)		1.7

*References 21, 24, 76, 108, 142, 155, 186.
†Reference 170.
‡References 158, 186.

throughout the membrane, data which demonstrate that the convex portion of the membrane may be richer in cholesterol than the concave portion exist.[150] Physical changes within the membrane which might correspond to such chemical changes have not, however, been demonstrated.[137] Mature red cells lack the ability to synthesize cholesterol from acetate.[96,148]

Using C^{14}—labeled lipoproteins obtained from dogs, Hagerman and Gould demonstrated that the cholesterol of red cell membranes was in an equilibrium exchange with the free cholesterol bound to serum lipoproteins.[61] This observation was confirmed for man in later studies, using either whole plasma or isolated alpha or beta lipoproteins in vitro[2,9,104] and deuterium[92] or C^{14}—labeled cholesterol in vivo.[58,86,136,168] Complete equilibration between plasma and cell free cholesterol occurred in approximately eight hours, both in vitro and in vivo. With the exception of one study which indicated that a portion of the free cholesterol of alpha lipoproteins did not participate in this exchange,[2] the equilibrium has appeared to involve the entire free cholesterol pool of both plasma and red cells. The rate of exchange in vivo has not been influenced by the presence of the nephrotic syndrome, xanthomatosis tuberosum or myxedema.[86,92] In normal subjects, however, the rate was more rapid when diets were rich in unsaturated fatty acids and slower when diets were rich in saturated fatty acids.[168] This exchange showed little dependence on temperature,[16] did not require energy,[104] and was not influenced by either neuraminidase treatment of red cell membranes or phospholipase D treatment of lipoproteins.[16]

Sterols other than cholesterol are capable of exchanging with cholesterol in the red cell membrane.[15] Willmer[195] has suggested that the action of steroid hormones may be related to their direct incorporation into membranes, and this may be true for progesterone.[32,33] Conflicting data, however, exist, and it is difficult at present to draw general conclusions.[59,129]

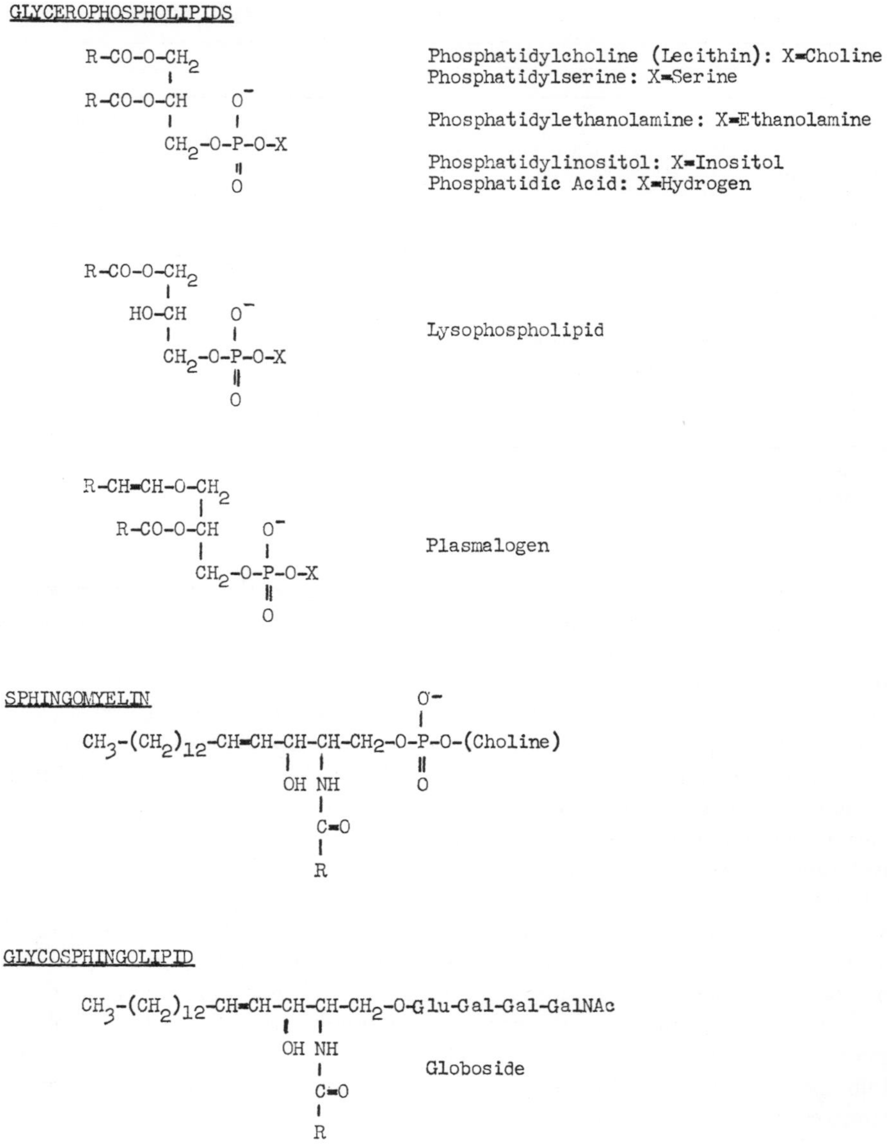

GLYCEROPHOSPHOLIPIDS

R-CO-O-CH₂
|
R-CO-O-CH   O⁻
| |
CH₂-O-P-O-X
‖
O

Phosphatidylcholine (Lecithin): X=Choline
Phosphatidylserine: X=Serine

Phosphatidylethanolamine: X=Ethanolamine

Phosphatidylinositol: X=Inositol
Phosphatidic Acid: X=Hydrogen

R-CO-O-CH₂
|
HO-CH   O⁻
| |
CH₂-O-P-O-X
‖
O

Lysophospholipid

R-CH=CH-O-CH₂
|
R-CO-O-CH   O⁻
| |
CH₂-O-P-O-X
‖
O

Plasmalogen

SPHINGOMYELIN

CH₃-(CH₂)₁₂-CH=CH-CH-CH-CH₂-O-P-O-(Choline)
                   | | ‖
                  OH NH   O
                     |
                     C=O
                     |
                     R

O⁻

GLYCOSPHINGOLIPID

CH₃-(CH₂)₁₂-CH=CH-CH-CH-CH₂-O-Glu-Gal-Gal-GalNAc
                   | |
                  OH NH
                     | Globoside
                     C=O
                     |
                     R

Fig. 1.

Phospholipids

Both glycerophospholipids and sphinogomyelin are present in the red cell membrane (Fig 1). The majority of glycerophospholipids contain a phosphoric acid ester attached to the third carbon and long chain fatty acids attached to the first and second carbons. In some phospholipids, however, one fatty acid is replaced by a vinyl ether (plasmalogen), and in some no fatty acid exists on the first or second carbon (lysophospholipid). Phosphatidic acid contains

Table 2.—Predominant Fatty Acids of Phospholipids in Normal Human Red Cells*

Fatty Acid	Phosphatidyl-ethanolamine %	Phosphatidyl-serine %	Lecithin %	Sphingomyelin %
16:0	14.2	3.6	33.0	32.4
18:0	12.8	38.6	12.8	7.9
18:1	17.6	8.1	20.0	2.2
18:2	6.4	2.9	22.4	2.0
20:4	22.7	23.8	6.7	0.8
22:0	0.8	1.0	1.3	8.7
22:5 and 22:6	13.0	11.5	1.3	1.0
24:0 and 24.1	0.6	6.2	0.3	38.6

*References 37 and 186.

a free phosphate group, rather than a phosphoric acid ester. Glycerophospholipids are subclassified according to the composition of their phosphoric acid ester: phosphatidylcholine (lecithin), phosphatidyl ethanolamine (PE), phosphatidyl serine (PS), and phosphatidyl inositol (PI). Sphingomyelin (SM) does not contain glycerol, but rather is composed of sphingosine and a fatty acid linked by a phosphoric acid ester to choline. The variety of phospholipid classes and the wide range of fatty acids and vinyl ethers that each may contain makes the number of structural possibilities almost infinite. Moreover, the option of a cis or trans configuration around the double bonds of fatty acids introduces more structural variability. Diacyl forms predominate among the glycerophospholipids in human red cells. However, 35 to 67 per cent of PE and 2 to 10 per cent of lecithin exist as plasmalogens.[42,63] This distribution varies among species, and vinyl ethers are not found in porcine or bovine red cell phospholipids.[63] The majority of PE in bovine red cells occurs as a glyceryl ether, a form not found in human red cells.[63] The predominant fatty acids present in the major phospholipid classes of the human red cell are shown in Table 2. Lecithin is characterized by the shorter, saturated fatty acids and SM by saturated fatty acids of chain length greater than 20. More than half of the red cell 18:0 resides in PS, and in both PS and PE greater than one-third of the fatty acids contain four to six double bonds. Because these polyunsaturated fatty acids are particularly labile to autoxidation, care must be exercised in the preparation of red cells for lipid analysis.[34,36,132]

Mature red cells are unable to synthesize fatty acids de novo, although they are able to bring about chain lengthening.[135] However, plasma free fatty acid (FFA) rapidly exchanges with FFA of red cell membranes.[158] Red cell FFA is actively incorporated into lysophosphatides utilizing ATP and coenzyme A (CoA),[125,158] the fatty acids so incorporated most commonly appearing in the 2-position of glycerol.[102,144,175,181] Studies in vitro have demonstrated different rates of incorporation for various fatty acids.[38,101] Since lysophospholipids also rapidly exchange between plasma and red cell membranes, the acylation of lysophospholipids within the membrane provides one mechanism for the renewal of membrane phospholipids which have been degraded.

A second mechanism of phospholipid renewal utilizing lysophospholipids, but not requiring energy, occurs by a dismutation of lysolecithin resulting in the generation of lecithin and glycerophosphorylcholine.[102,175] Both the energy-requiring acylation and the dismutation of lysophospholipid involve predominantly lecithin. Fatty acid incorporation into PE has also been demonstrated but to a lesser extent. PE renewal however, involves not only the direct acylation of lyso-PE by FFA, but also an indirect transacylation in which the fatty acid donor is lecithin, lysolecithin being generated in the process.[156] Carnitine acyl-transferase activity has also been found in red cell membranes, but it is not essential for the acylation of lysophosphatides described above and its role in the membrane is unclear.[99]

Phospholipid renewal within red cell membranes may also occur by a direct exchange of the intact molecule with its counterpart in plasma.[139,175] Studies in vitro utilizing red cell and plasma phospholipids labeled with P^{32} in vivo have demonstrated that both lecithin and SM participate in this exchange, resulting in a 9 per cent turnover of lecithin and a 4 per cent turnover of SM in 12 hours.[139] Exchange data suggest that these two phospholipids are not homogeneous in the red cell membrane, but rather that only a portion of each is exchangeable, amounting to 60 per cent of lecithin and 30 per cent of SM in human red cells.[139] Qualitatively similar results have been obtained with canine and rat red cells.[90,139]

The turnover of red cell lecithin by the three mechanisms discussed above was studied in rat red cells.[175] The total turnover of lecithin was 5.8 per cent in three hours: 1.2 per cent resulted from the acylation of lysolecithin utilizing fatty acids, ATP, and CoA; 1.6 per cent resulted from the dismutation (or condensation) of two molecules of lysolecithin; and 3 per cent resulted from a direct exchange with plasma lecithin.

Controversy exists regarding the turnover of phospholipid phosphorous in mature red cells, as studied with P^{32}—labeled orthophosphate. All workers agree that the phosphate of phosphatidic acid, a phospholipid present in trace amounts in red cell membranes, turns over, deriving its phosphate from ATP. Turnover of the phosphate group in phosphatidyl inositol has also been demonstrated.[69] However, the incorporation of P^{32}—orthophosphate into other phosphatides, PS[74,147] and PE[60] in particular, has been disputed.[138,191] The total turnover of phospholipid phosphorous in vitro is exceedingly slow and accounts for less than 0.1 per cent of the total membrane lipid phosphorous per hour.[138]

Glycolipids

Glycolipids are complex molecules containing sphingosine, a long chain fatty acid, and one or more molecules of hexose or hexosamine (Fig. 1). The presence of sphingosine and a fatty acid makes this class of compounds similar to SM. Indeed, like SM, the fatty acids are predominantly 24:0 and 24:1.[199a] Sphingosine and fatty acid together are referred to as ceramide.

The glycolipids of mammals can be divided into two major classes. Those which contain glucose, galactose and N-acetylgalactosamine (GalNAc) or

N-acetylglucosamine (GluNAc) are termed globosides, and those which contain glucose, galactose and N-acetylneuraminic acid (NANA) or N-glycolylneuraminic acid (NGNA) are termed hematosides. Human red cell membranes contain glycolipids of the globoside (GL) type with one to four (GL-1 to GL-4) hexose residues.[170] The number of hexose residues per glycolipid varies: up to three in dogs, cats and horses, four in humans and most other mammals, and five in rabbits. In each species, glycolipids with the maximum number of hexose residues predominate. It is of interest that in pig red cells similar fatty acids are present in GL-3 and GL-4, but these differ from the fatty acids present in GL-1 and GL-2.[170] Minor glycolipids are also present, in particular those with A, B, and Lewis blood group activity.[62]

The stability of red cell membrane glycolipids in vivo has been studied in the pig[170] and the rabbit.[85] Following the injection of C^{14}–labeled glucose, membrane GL-1 rapidly exchanges with plasma, primarily with low density lipoproteins. GL-4 in the pig and GL-5 in the rabbit do not undergo exchange, labeled GL-4 and GL-5 of red cell origin appearing in plasma only at the time of red cell destruction. GL-2 and GL-3 appear to behave in an intermediate way with 40 to 60 per cent of the labeled molecules being lost from the membrane over the first few days, indicating exchange with plasma, and the remainder being released at the time of red cell destruction. The Lewis blood group glycolipids, complex lipids containing glucose, galactose, fucose, and GluNAc, are transported by plasma high and low density lipoproteins and are in exchange equilibrium with the red cell membrane.[95]

It appears that human red cell glycolipids contain very little[14a] or no[198] neuraminic acid and that the neuraminic acid present in red cells is confined to glycoproteins. The relative amounts of neutral and amino sugars in human red cell glycoproteins and glycolipids are approximately equal.[198] Similarities in the sugar sequence make it likely that certain antigenic properties are shared by both glycolipids and glycoproteins. Glycolipids are not found in mitochondria or membranes of the endoplasmic reticulum and they have not been found in nuclei.[46,47] Rather, it appears that this important class of lipids is unique to the plasma membrane.

PHYSIOLOGIC ALTERATIONS IN RED CELL LIPIDS

Dietary Changes

Since red cell lipids are, to a greater or lesser extent, in equilibrium exchange with plasma lipids, and since plasma lipids undergo striking changes under a variety of dietary conditions, one might expect that red cell lipids would follow these dietary changes. Although changes do occur in red cell fatty acid composition, there is no change in the red cell content of cholesterol or phospholipid in man[43] or rodents[183] treated with a fat free diet or in man treated with diets rich in various triglycerides.[43] In patients on a diet free of fat, red cell 18:0 and 18:1 increase and 18:2 reciprocally decreases, reflecting the fatty acids synthesized from carbohydrate.[43] Although these changes in red cell fatty acids reflect changes in plasma, the fatty acid composition of plasma changes within several days, whereas changes in red cells occur over the course of weeks. With diets rich in corn oil, red cell 18:2 increases and, depending on the amount of corn oil, this reaches a maximum in six weeks to six months. Changes occur in all phospholipid classes, including the fatty aldehydes of PE plasmalogen. In general, the content of 18:2 in red cell membranes correlates closely with its content in the diet, since 18:2 is not synthesized by mammals, and red cell 18:1 and 18:2 fluctuate in a reciprocal fashion.[43]

Rats maintained on a fat free diet undergo similar red cell changes: there is a decrease in 18:2 and 20:4, an increase in 16:1 and 18:1, and the appearance of 20:3.[41,182] Lecithin increases and PE and PS decrease.[41] Upon fat refeeding, 20:3 decreases and 20:4 increases.[115] The effect of dietary change on red cell survival has not been studied in man. On diets free of fat, rats have red cells with an increased osmotic fragility, an altered appearance on electron microscopy, and a decreased survival.[41,97]

Red Cell Aging In Vivo

When released from the bone marrow as reticulocytes, red cells are particularly rich in membrane lipid. In rats, the lipid content of young reticulocytes is 60-80 per cent greater per cell than that of mature red cells.[20a] During the first two to three days after release, reticulocytes are remodeled into mature cells. During this remodeling, there is a loss of reticulum, water, volume and membrane lipid.[20a] Attempts to demonstrate age dependent changes in mature red cells have been made using density gradient centrifugation to separate cells of various ages. Young red cells are more buoyant and the top fraction, therefore, reflects differences known to be present in reticulocytes. The bottom fractions contain a mixture of cells which tend to be older, but this includes cells of any age which are smaller and more dense. The lipid content per red cell in the bottom fraction is 5 to 10 per cent less than the entire cell population.[180,192] The phospholipid composition, however, is unchanged.[180] Whether this observation demonstrates that older cells have a lipid content which is less than the whole cell population or that cells which are more dense regardless of age have a lipid content less than cells of average density is unclear. No concrete evidence exists demonstrating that, following maturation, red cells continue to lose lipid.

Changes in Red Cell Lipids During Storage

Normal red cells stored at refrigerator temperatures in acid-citrate-dextrose (ACD) undergo a progressive loss of both cholesterol and phospholipid, amounting to approximately 25 per cent in seven weeks.[64,140] Half of this loss occurs within the first two weeks. The loss of phospholipid affects all of the major phosphatides, and no change in their relative amounts is evident after eight weeks of storage. A decrease in critical hemolytic volume (an indirect measure of surface area) accompanies this lipid loss.[64] In contrast to this symmetrical los of lipid, saline-washed red cells frozen in saline containing EDTA undergo a progressive, but selective, loss of glycerophosphatides during storage.[185] This alteration results in a decrease of total lipid phosphorus by more than 30 per cent after eight weeks of storage. There is little or no change in the red cell content of SM or cholesterol.

Changes in Red Cell Lipids In Vitro

During incubation in vitro, the lipids of normal red cells equilibrate with plasma lipids to varying degrees, as discussed above. This equilibration, however, does not cause quantitative changes in red cell lipids. Changes in the lipid content of red cells in vitro have been observed under two sets of conditions:[25] First, when incubated in serum under homeostatic conditions with respect to glucose, red cell membranes undergo a selective loss of cholesterol;[103] and second, in contrast, when incubated under conditions leading to metabolic depletion, both cholesterol and phospholipid are lost in a symmetrical fashion.[141]

The selective loss of cholesterol from red cells incubated in serum results from the esterification of free cholesterol in serum by the serum enzyme, lecithin:cholesterol acyltransferase (LCAT). This enzyme was first described by Sperry in 1935[167] but was largely ignored until the recent studies by Glomset[56] and others. LCAT transfers a fatty acid from the beta– (or 2–) position of lecithin to cholesterol, thus forming a cholesteryl ester. Since cholesteryl esters do not form a part of the red cell membrane, the esterification of cholesterol decreases the amount of cholesterol in the serum-cell pool. The cholesteryl esters so formed apparently occupy binding sities within serum lipoproteins different from the sites occupied by the free cholesterol from which they were formed. Following the action of LCAT, serum lipoproteins have an increased affinity for free cholesterol.[26] As shown by Murphy[103] and confirmed by others,[15,22,25] when red cells are incubated in serum partially depleted of free cholesterol by the action of LCAT, cholesterol is lost from red cells into serum lipoproteins. Red cells are not unique in this regard, and serum depleted in this manner will accept cholesterol from platelets as well.[20a] The loss of cholesterol from membranes is associated with a proportional loss of surface area, as measured indirectly by osmotic fragility.[22,103] Red cells depleted of free cholesterol rapidly regain both cholesterol[25,103] and surface area[25] upon reincubation in serum which has been heated to 56°C to destroy LCAT activity. Moreover, red cells made osmotically fragile by cholesterol depletion in vitro rapidly acquire a normal osmotic fragility following reinfusion in vivo and survive normally.[25] Although the exact role of LCAT in vivo is unknown, evidence has accumulated in recent years that would ascribe to it a major role in the production of serum cholesteryl esters.[56,116,117] A congenital deficiency of this enzyme is associated with very low levels of cholesteryl esters in serum.[119]

When incubated under conditions leading to metabolic depletion, red cells undergo a loss of *both* cholesterol and phospholipid. This change occurs prior to the onset of measurable hemolysis.[25,141] Similar lipid loss has been observed with red cell ghosts.[88] Protein is not lost from ghosts under these conditions, suggesting that lipid loss occurs at least in part, because of an altered membrane affinity for lipid.[88] The late loss of lipid from intact red cells, however, is accompanied by a loss of membrane protein.[153a] Associated with this lipid loss is a proportional decrease in membrane surface area, as estimated from critical hemolytic volume.[188,189] In contrast to the selective loss of cholesterol, the loss of lipid under these conditions is irreversible.[25]

CHANGES IN RED CELL LIPIDS IN DISEASE

Hereditary Spherocytosis

The lipid content of red cells from patients with hereditary spherocytosis (HS) and intact spleens is usually normal or decreased,[24,89,141] despite the presence of increased numbers of reticulocytes. The relative amounts of cholesterol and phospholipid and various phosphatides are normal,[24,30,133,141] although an earlier report had indicated a decreased amount of PE in HS cells.[80] Following splenectomy in HS, reticulocytes fall to normal or near normal levels and red cell lipids are the same as in normal subjects with intact spleens.[14,24,133,141] However, splenectomy itself influences the red cell membrane. After splenectomy in patients without HS, target cells with increased surface area occur and, parallel with their appearance, there is an increase in osmotic resistance.[8] Both evolve to a maximum over several weeks to several months.[100,164,178] The red cell content of lipid is increased 15 to 20 per cent in patients without HS who have undergone splenectomy, as compared with normal subjects with intact spleens.[24] Comparison of red cells from patients with and without HS, all of whom have undergone splenectomy, reveals, therefore, a 15 to 20 per cent deficiency of both cholesterol and phospholipid in the HS cells.[24] HS cells after splenectomy, however, have a lipid content which is, nonetheless, similar to that of red cells from normal subjects whose spleens are intact. Despite this similarity, red cells from HS patients without spleens are more osmotically fragile than red cells from normal subjects with spleens. This difference indicates that factors in addition to lipid deficiency per se are responsible for the deficiency of surface area in HS red cells.

When incubated under conditions of glycolytic homeostasis, HS red cells undergo normal changes in that they selectively and reversibly lose cholesterol.[24] However, under conditions of metabolic depletion, HS red cells lose both cholesterol and phospholipid in amounts which are far in excess of normal.[24,89,141] Thus, in addition to a decreased surface area and an increased permeability to sodium,[12,73] the membrane in HS red cells is deficient in lipid, both cholesterol and phospholipid, and its lipids are particularly unstable, undergoing inordinate loss in the face of energy depletion. In bovine red cells, membrane stability depends upon the presence of divalent cations.[18] Recent data indicate that an abnormal interation between ATP, divalent cations, and the membrane may exist in HS red cells, and that this abnormality may account for the cells' leakiness to sodium and instability upon ATP depletion.[87]

It has been suggested that the loss of lipid in vitro by HS red cells is related to an increased phosphatide turnover which accompanies an increased rate of active cation pumping.[72] It appears, however, that lipid loss in HS red cells can be partially or completely prevented by maintaining a physiologic pH and adequate levels of glucose, despite this increased sodium flux.[24,141] Moreover, an increased red cell sodium flux in other conditions is not associated with a loss of membrane lipid in vitro.[127,200] No correlation exists between the sodium and potassium content of red cells in various animal species and the lipid composition of their membranes,[111] and no differences have been found in the lipid composition in red cells obtained from high potassium (HK) and low potassium (LK) sheep.[113,114] Thus, cation pumping does not appear to cause lipid loss from red cell membranes. Evidence that membrane sterols do not play an important role in sodium permeability is indicated by a lack of change in the sodium flux of normal red cells depleted of cholesterol by LCAT[103] or of HS red cells which have acquired excess cholesterol in vitro.[24]

In addition to an increased sodium flux, red cells in HS have an increase in the activity of the Na^+-K^+ responsive ATPase.[106,194] Moreover, a close correlation exists between the rate of sodium flux in intact cells and the activity of the Na^+-K^+ ATPase in ghosts prepared from the same red cells.[194] However, sodium flux does not correlate with survival of HS red cells in vivo.[193a] Several studies have been carried out in an attempt to define the relationship between this membrane enzyme and membrane lipids. Extraction of red cell ghosts with ether to remove the loosely bound lipids does not influence ATPase activity.[145] Sodium dodecyl sulfate in low concentrations increases its activity and low concentrations of deoxycholate have no effect: however, high concentrations of both of these lipid active substances inhibit the enzyme.[19,165] Phospholipases A and C, lysolecithin, and n-butanol all decrease its activity.[145,165] The addition of phospholipid to the solubilized ATPase appears to be essential for its activity.[45] Although lipid appears to be important in the Na^+-K^+ ATPase, the relation between this enzyme and the lipid abnormalities in the HS cells is unclear.

Thus, despite considerable data regarding membrane lipids in HS and the role of lipids in various membrane phenomona which appear abnormal in HS, the nature of the membrane defect in this interesting disorder is still unknown. However, the acquisition of lipid and surface area by HS red cells in vivo appears to circumvent this defect and to allow a marked prolongation of cell survival.[24]

Other Hemolytic Anemias

In contrast to the extensive data in HS, few data exist regarding the red cell lipid changes in spherocytosis induced by antibodies. The red cells of patients with gamma-Coombs positive hemolytic anemia are deficient in both cholesterol and phospholipid (averaging approximately 10 per cent), despite the presence of increased numbers of reticulocytes.[13] Similarly, in rats the spherocytosis which occurs progressively over the course of 20 hours following the injection of heterologous red cell antibodies is associated with a 20 to 30 per cent loss of cholesterol and phospholipid and a 6 per cent loss of protein from the red cell membrane.[20] Thus, membrane lipid loss appears to play a role in the pathogenesis of immune spherocytosis. The mechanism of this loss awaits further studies.

A family has recently been described, eight members of which have a hemolytic anemia and red cells containing a 25 per cent increase in lecithin content, but a normal content of cholesterol and total phospholipid.[76] Red cells from affected patients have a decreased ability to transacylate membrane bound lyso-PE by membrane bound lecithin, possibly accounting for the accumulation of lecithin in this disorder.[157] The lipid disturbance, however, appears subtle, and its role in the hemolysis has not been defined.

Sparse observations exist in other hemolytic anemias. An increase in membrane cholesterol, phospholipid and surface area of approximately 15 per cent has been observed in red cells in sickle cell anemia when compared with the young fraction of cells separated by centrifugation from normal blood.[193] The phospholipid composition of red cells in paroxysmal nocturnal hemoglobinuria (PNH) was initially described as abnormal,[65] but subsequent studies have established that the phospholipids in this disorder are normal.[14,133] The cholesterol and phospholipid in glucose-6-phosphate dehydrogenase (G-6-PD) deficient red cells are normal.[174] Incubation of normal red cells with primaquine, an agent which induces hemolysis in vivo in G-6-PD deficiency, results in an increased incorporation of C^{14}—labeled fatty acids into lecithin, indicating an increase in the rate of lysolecithin generation and its reacylation to lecithin.[199] Additional studies are required in these and other hemolytic anemias.

Liver Disease

Patients with various forms of liver disease, including hepatitis, cirrhosis, and obstructive jaundice, have red cells with an increased content of both cholesterol and phospholipid.[17,22,108,122,190] The cholesterol increase may be as great as 75 per cent but more commonly ranges between 25 and 50 per cent greater than normal. Although there is variability from patient to patient, the percent increase in phospholipid is approximately one half of the percent increase in cholesterol, resulting in an increased cholesterol: phospholipid ratio. In four separate studies, this averaged 18 per cent (Table 3). The phospholipid increase is not distributed among the various phospholipids but rather is confined to lecithin.[108,122,184,190] Thus, cholesterol and lecithin are not only the most exchangeable of the major red cell lipids, as discussed above, but their membrane compartments also undergo the greatest quantitative changes in liver disease. Because the percent increase in lecithin exceeds that of cholesterol, the cholesterol:lecithin ratio decreases (Table 3). While changes are seen in most patients with liver disease, they are most striking in patients with obstructive jaundice or with hepatitis and an obstructive component. One might expect the changes in red cells to be a direct reflection of changes in serum lipids, since the serum concentrations of cholesterol and lecithin also increase. However, no correlation has been found between the red cell content and the serum concentrations of either cholesterol or phospholipid. In fact, elevations in the red cell content of lipid are

Table 3.—Variations in Red Cell Cholesterol and Phospholipid in Disease

References	Normal (All)	Target Cells in Liver Disease (20a,23,108,190)	Spur Cells in Liver Disease (20a,21)	Acantho-cytosis (20a,155,187)	LCAT Deficiency (20a,55)
Cholesterol $\mu g/10^8$ Cells	12.24	18.10	20.90	12.68	18.30
Phospholipid $\mu g/10^8$ Cells	29.58	36.98	32.15	27.10	25.32
Cholesterol: Phospholipid Mole:Mole	0.83	0.98	1.30	0.94	1.44
Lecithin* %	32.0	45.2	35.8	21.0	57.3
Lecithin† $\mu g/10^8$ Cells	9.46	16.71	11.51	5.70	14.51
Cholesterol: Lecithin Mole:Mole	2.59	2.17	3.64	4.46	2.53

*The per cent lecithin in normal controls varies among the reports. It was normalized to 32 per cent and the per cent lecithin was recalculated for patients using the following formula:

$$\frac{\% \text{ lecithin patients}}{\% \text{ lecithin controls}} \times 32\%$$

†Per cent lecithin. $\times$ reported value for total phospholipid.

found even in patients whose serum concentrations of cholesterol and phospholipid are less than normal.[17,20a,108]

Increases in membrane cholesterol content correlate with increases in membrane surface area, as measured by osmotic fragility.[17] Extension of their surface area causes red cells to become broad and flat and to appear cup-shaped in wet preparations and targeted on dried smears.[8] These changes in the lipid content, osmotic fragility and appearance of red cells are reversible within days to weeks following the relief of biliary obstruction or the subsidence of intrahepatic disease.[11,17,53,66,110,190] Using Cr^{51}, it has been found that normal red cells progressively acquire surface area and become resistant to osmotic lysis in the circulation of patients with obstructive jaundice over the course of two to three days, and that osmotically resistant target cells lose surface area and become osmotically normal over a similar time course in the circulation of normal subjects.[22] These changes in vivo have been reproduced in vitro using normal red cells and sera from patients with obstructive jaundice.[22] Under these conditions, normal red cells acquire cholesterol and become resistant to osmotic lysis but undergo no change in their content of phospholipid. Thus, this red cell lipid abnormality is acquired and reversible and dependent upon serum factors. The lack of a measurable change in red cell lecithin in vitro may be related to the relatively slow rate of exchange between serum and cell lecithin, in contrast to the rapid equilibrium between serum and cell free cholesterol. Whether or not cholesterol and phospholipid are acquired by the same mechanism, it is clear that cholesterol acquisition may occur independent of any change in phospholipid.

The mechanism of lipid acquisition in liver disease has not been completely defined, but several factors appear to be important.[22] First, the activity of the serum enzyme, LCAT, is decreased in patients with obstructive jaundice.[22,179] Since free cholesterol and lecithin are the substrates for this enzyme, a decrease in its activity would be expected to favor the accumulation of these lipids in plasma. Thus, if not causing red cell lipid accumulation itself, the relative lack of this enzyme serves a permissive role in assuring the availability in serum of those lipids which accumulate in the red cell. A second factor is the elevated serum concentration of bile salts in patients with liver disease, obstructive jaundice in particular.[22] This accumulation may, in fact, be partially responsible for the depression of LCAT activity.[22,167] An additional, and possibly more important, effect of bile salts is their influence on the equilibrium partition of free cholesterol between serum lipoproteins and red cell membranes.[22] When added to normal serum containing normal red cells, bile salts cause the acquisition by these red cells of both cholesterol and surface area. Whether bile salts act primarily on red cells or on serum lipoproteins is not clear.

Patients with obstructive jaundice have an immunologically distinct low density lipoprotein,[149] characterized originally as the obstruction lipoprotein (OLP)[171] and more recently renamed LP-X.[1,153] This lipoprotein is particularly rich in free cholesterol and appears to be relatively distinct for obstructive jaundice, occurring in a very high percentage of patients with extrahepatic biliary obstruction and biliary cirrhosis and some patients with the cholestatic syndrome and Laennec's cirrhosis. It is found only rarely in infectious or serum hepatitis.[153,172] The exact relationship between this lipoprotein and red cell lipids is unknown. The fact that striking changes are seen in red cell lipids in hepatitis,[17] whereas this abnormal lipoprotein is rarely found in this disorder, makes a direct association appear unlikely.

The red cell lipid abnormalities in liver disease are not necessarily associated with a shortening of red cell survival, although patients with alcoholic fatty livers may have transient periods of hemolysis coincident with their active disease, and patients with cirrhosis may have a chronic hemolytic anemia.[77,83,201,202] Occasionally, patients with alcoholism and fatty livers have a gross elevation of their serum triglycerides. This increase may result from several known factors: first, a decreased activity of postheparin lipolytic activity (PHILA);[93] second, the induction of endogenous lipemia by carbohydrate;[106] and third, pancreatitis.[82] In the absence of alcoholism and a fatty liver, hypertriglyceridemia is not known to be associated with hemolysis.[3] The triad of hyperlipemia, fatty liver and hemolytic anemia has been grouped into a disease entity known as Zieve's syndrome.[83,190,201] Red cell lipids in this "syndrome" are typical of those in patients with liver disease in general; i.e., there is an increase in cholesterol of approximately 50 per cent and of phospholipid of approximately 25 per cent greater than normal, lecithin being the major phospholipid which is increased.[190] While the young alcoholic with milky serum, a large, tender fatty liver and hemolytic anemia presents a striking picture, the various features of this picture may not be related pathogenetically.

Over the past five years, a small number of patients has been reported with hemolytic anemia and red cells resembling acanthocytes in association with severe hepatocellular disease.[26,159,166] In six patients studied in our laboratory, the cholesterol content of the so-called "spur" red cells in this disorder has been increased by 40 to 70 per cent with little or no change in the red cell content of phospholipid.[20a,21] The cholesterol:phospholipid ratio is increased, therefore, by as much as 75 per cent (Table 3). In contrast to target cells, the lecithin content of spur cells is normal or near normal, and the cholesterol:lecithin ratio is markedly increased.[20a] Cholesterol accumulation results from the passive, reversible acquisition of free cholesterol from serum lipoproteins.[21] Cholesterol-loaded spur cells have a low affinity for this excess free cholesterol and transfer it to normal serum lipoproteins in vitro. However, the normal serum which has gained cholesterol by this means holds it with a low affinity and readily transfers it to normal red cells incubated therein.[26] Coincidental with the acquisition of cholesterol in vitro, membrane architecture is distorted, causing normal red cells to become spiculated in appearance. These changes in the red cell membrane impede the cell in its circulation in vivo and result in a shortened cell survival. Studies in vitro and in vivo have shown an increase in membrane surface area accompanying the acquisition of cholesterol. However, this increase does not persist in vivo, membrane loss being a feature of the red cell destruction in this disorder.[21]

Why some patients with liver disease develop spur cells and others have target cells is unclear. The cholesterol:phospholipid ratio is increased in both, although to a greater degree in spur cells than in target cells. The cholesterol:lecithin ratio, however, is markedly increased in spur cells, whereas it is decreased in target cells. Spur cells occur in severe hepatocellular disease, a condition in which bile acid metabolism is diverted to the production of chenodeoxycholic acid and its bacterial degradation product, lithocholic acid. In contrast, patients with obstructive jaundice accumulate cholic and deoxycholic acids. Rhesus monkeys tolerate the addition of cholic acid to their diets, but, when fed lithocholic acid, they develop cholesterol loaded, spur-shaped red cells over the course of days.[21a]

Lipoprotein Disorders

Hyperlipoproteinemia. Red cell cholesterol and phospholipid are normal in patients with familial hypercholesterolemia (Type II).[21a,44,49] Patients with hypertriglyceridemia due to increases in chylomicrons (Type I) have red cells with a normal or moderately decreased content of cholesterol and phospholipid.[20a,44,107] Similarly, in endogenous hypertriglyceridemia (Type

IV), red cell cholesterol and phospholipid are normal[107] or slightly reduced.[3] Red cell survivals are normal in these disorders.[3] The total number of patients with hyperlipoproteinemia reported is small, and more studies are needed before firm conclusions can be made. However, from the available data, it appears that gross changes in serum lipoproteins may exist with little or no change in red cell lipids.

Alpha Lipoprotein Deficiency (Tangier Disease). Several kindreds have been described with a deficiency in serum lipids due to the congenital absence of normal alpha lipoprotein.[51,68] A small amount of alpha-like protein persists in their serum. In five reported cases, the red cells in this disorder have had a normal content of both cholesterol and phospholipid; however, in each case the percent lecithin has been greater than normal.[50,155] In one patient, the red cell fatty acids contain an increased quantity of 18:1 and 18:0 and a decreased quantity of 16:1 and 18:2. The fatty acid distribution, however, became normal when this patient was placed on a low carbohydrate diet.[155]

Beta Lipoprotein Deficiency (Acanthocytosis). Acanthocytosis is a rare disease resulting from the congenital absence of beta lipoprotein. Fewer than 30 cases have been reported. It takes its name from the thorny appearance of red cells, as first observed by Bassen and Kornzweig.[10] The plasma level of triglycerides approaches zero, cholesterol is usually less than 50 mg. per 100 ml. and phospholipids less than 100 mg. per 100 ml.[91,187] SM accounts for a greater than normal proportion of plasma phospholipids and there is a reciprocal decrease in the percent of lecithin.[78,187] The fatty acids esterified to plasma phospholipids contain one-third to one-half the normal content of 18:2.[87]

The cholesterol content of acanthocytes is normal[20a,50,131,187] or slightly increased,[91,98] and the phospholipid content is normal or slightly decreased. Red cell phospholipid and fatty acid composition reflect that of serum.[187] Although variability exists in the reported cases, cholesterol:phospholipid ratios tend to be increased (Table 3). Red cell lecithin is decreased and SM increased, and the cholesterol:lecithin ratio is increased to levels also seen in spur cells. Attempts to induce or reverse the red cell morphologic abnormality in vitro have been unsuccessful;[151,152] however, at a time when 80 per cent of the red cell mass in a patient with acanthocytosis was normal transfused blood, greater than 50 per cent of the red cells were altered in appearance.[52] The morphologic features persist after washing in saline and resuspension in 20 per cent albumin[163] but are reversed by nonionic (although not by ionic) detergents.[173] Autohemolysis[81,162] and peroxidative hemolysis[34] are increased and are corrected by the addition of vitamin E, either in vitro or in vivo.[34,81] Vitamin E does not, however, alter the lipid composition or morphology of acanthocytes, and the requirement for this fat-soluble vitamin reflects a deficiency state induced by malabsorption. Anemia, when present, is usually mild, and the t½ of Cr[51]–labeled cells is only slightly shortened.[162,187] Reticulocytes may be normal or slightly increased in number and the osmotic fragility is usually within the normal range.[162] The morphology of red cells in this disorder is indistinguishable from spur cells in liver disease; in both, the cholesterol:lecithin ratio is markedly increased. Whether or not the changes in membrane lipid composition cause these changes in membrane architecture is unclear.

LCAT Deficiency. A deficiency in alpha lipoprotein associated with the absence of measurable LCAT in plasma has been reported in a kindred of three adult sisters.[54,177] No other family members had a demonstrable lipid abnormality. All patients had a marked increase in the serum level of free cholesterol, very low levels of cholesteryl esters, elevated lecithin and, in two, elevated triglycerides. The clinical syndrome included moderate anemia with a hemolytic component, proteinuria and hypoalbuminemia. The liver was histologically normal, except for increased amounts of glycogen, and both the liver function tests and serum creatinine were normal. Multiple abnormalities in plasma lipoproteins have been noted including: three abnormal lipoprotein bands on electrophoresis which were reactive with antiserum against normal alpha lipoproteins; an abnormal beta lipoprotein which sedimented with the high density lipoproteins; and an abnormal very low density lipoprotein (VLDL) which, despite increased serum triglyceride levels, formed no pre-beta band on electrophoresis.[118,176] The small amounts of cholesteryl esters found in plasma were derived from intestinal absorption.[120] Red cells in affected patients were targeted in appearance and contained a 50 to 80 per cent increase in cholesterol content[55] (Table 3). Red cell phospholipids were normal in total amount but lecithin accounted for more than 50 per cent of the phospholipid, and PE and SM were decreased in amount. Corresponding to this shift in phospholipids, 18:2 was increased and 20:4 decreased. Cholesterol-loaded red cells in this disorder lost cholesterol into normal serum and normal cells gained cholesterol when incubated in serum of affected patients.[121]

Certain similarities exist between this disorder and patients with liver disease. In both, LCAT is deficient. In liver disease, however, the deficiency is often mild, with LCAT activity averaging 50 per cent of normal.[21,22] The cholesterol:phospholipid ratio of red cells in LCAT deficiency is markedly increased and resembles that of spur cells. The cholesterol:lecithin ratio, like that of target cells in obstructive jaundice, is less than normal (Table 3). Complex interrelationships must exist in these various disorders between LCAT deficiency, the lipid affinity of normal and abnormal lipoproteins, and the effect of normal and abnormal bile salts on cell:serum lipid partitions; these are still incompletely defined.

Lipoprotein Abnormalities in Experimental Animals. Guinea pigs[124,128] and rabbits[67,134,761] receiving diets containing one per cent cholesterol develop serum levels of cholesterol in excess of 2000 mg. per 100 ml. and a disproportionate amount is free cholesterol. Serum phospholipids increase in proportion to cholesterol. Several weeks to several months after this rise in serum lipids, hemolytic anemia with splenomegaly develops. Red cell survival is approximately one-half normal, and normal transfused cells acquire the hemolytic defect.[134] In rabbits, the red cells resemble acanthocytes in appearance.[161] The lipid content of red cells is increased in both species. In guinea pigs, there is a two-three fold increase in both cholesterol and phospholipid,[128] whereas in rabbits cholesterol alone is increased.[167] Up to half of the increase in cholesterol is in the form of cholesteryl esters and a portion of the lipid increase in rabbits is in intracellular vacuoles.[67] Cholesterol loads are handled very poorly by these animals and increased amounts of cholesteryl esters persist in the livers of guinea pigs four to six months after the cessation of cholesterol feeding.[123]

Beta lipoprotein deficiency has been produced experimentally in rats by feeding a diet containing one per cent orotic acid.[197] The red cells of orotic acid treated rats contain a 20 to 50 per cent increase in cholesterol content, but a normal content of phospholipid.[98] The cells' osmotic fragility is decreased and they are spiculated in appearance. Cr^{51} survivals are decreased by 35 per cent with red cell destruction occurring predominantly in the spleen.

While these experimental models may not be directly applicable to questions of red cell-lipoprotein interactions in man, they do demonstrate that, as in man, the lipid content of red cells in experimental animals can vary widely and that cholesterol and phospholipid may behave independent of one another in this variation.

Miscellaneous Disorders

Vitamin E Deficiency. Vitamin E is a lipid soluble antioxidant which is in exchange equilibrium between red cell membranes and serum lipoproteins.[160] In rats receiving vitamin E deficient diets, red cells are deficient in PE and PS.[34,75] This deficiency results from the preferential oxidation or peroxidation of these two phospholipids, probably due to their high content of polyunsaturated fatty acids.[36] Peroxidative hemolysis is also increased in hereditary acanthocytosis, and this is corrected by vitamin E in vitro or in vivo.[34] Indeed, lipid peroxidation due to chronic vitamin E deficiency in acanthocytosis may underly at least some of the progressive neurologic changes seen in this disorder. Hemolytic anemia due to vitamin E deficiency has not been documented satisfactorily in adults. Hemolysis in premature infants 6 to 10 weeks old, however, appears to be related to vitamin E deficiency and it is corrected by the parenteral administration of vitamin E.[126] In a prospective series of premature infants treated with vitamin E, hemolysis has been significantly less than in untreated controls.[126] Mechanisms for this apparent sensitivity to vitamin E deficiency in premature infants are unknown. Although the membrane content of cholesterol and phospholipid in cord blood red cells is increased as compared with adult red cells, the phospholipid fractions and the fatty acids esterified to phospholipids show only minor deviations from normal.[27,109] Thus, this sensitivity may reflect other as yet undescribed variations in the structure of fetal red cell membranes.

Other Disorders. Red cells in hypothyroidism have been reported to have an increased content of cholesterol and a decreased content of lecithin,[71] and a deficiency of PE, PS, and SM has been reported in the red cells of patients with Tay Sach's disease, Nieman Pick's disease and Gaucher's disease.[5,6,41] Red cells in Huntington's chorea have a two-fold increase in glycolipid-NANA, although the plasma content of glycolipid-NANA is normal.[70] While one cannot draw general conclusions from these random observations, when grouped with other data they will, hopefully, give insight into the relationships between structure and function in the red cell membrane.

Conclusions

The lipid composition of red cell membranes varies considerably between species;[29] however, within a species, red cell lipids are remarkably constant. Although changes in red cell fatty acids result from dietary manipulations, changes in other lipids in vivo occur only in disease. The range of changes is broad. Indeed, when one constructs models of the red cell membrane, not only must the normal lipid composition be considered, but the extreme variability in lipid composition under experimental and pathologic conditions must also be considered.

Early concepts of membrane structure envisioned a bimolecular lipid leaflet,[28] and this appeared to be confirmed by electron microscopy of thin sections.[143] However, this theory has been criticized[84] and defended,[169] and must be viewed as possible but not established. The surface area of red cell lipids, as estimated from Langmuir troughs, has been related to independent measurements of red cell surface area.[7,31,57] At very low surface pressures, the surface area of red cell lipids is great enough to permit a bilayer. Theoretical calculations based on the known geometry of individual membrane lipids have also been consistent with a bilayer.[39] However, the surface pressure within the membrane is unknown, and at high surface pressures the amount of lipid is insufficient for a bilamellar membrane. The recent discovery of nonlamellar, two-dimensional lecithin soaps permits the consideration of lipid leaflets which are not uniform, but rather are lattice-like and able to accommodate protein molecules.[94] Nor must all lipid molecules participate in a single structure: thus, evidence has been cited above for the existence of two compartments of lecithin and SM, one in exchange with plasma and one not. That lipid is important in determining surface area is shown in Fig. 2 which demonstrates the close correlation between red cell cholesterol content after incubation in various sera in vitro and red cell surface area, as measured by osmotic fragility.

Variations in lipid composition are often selective and involve a single lipid class. Under certain conditions, particularly those involving lipid loss, subunits of the membrane containing all lipid classes appear to be involved. With both antibodies and metabolic depletion, however, this conjoint lipid loss is accompanied by the loss of little or no membrane protein.

The effect of various lipids on the permeability of biological membranes has received little attention in comparison to the extensive work involving artificial lipid membranes. While changes in membrane cholesterol content do not appear to influence sodium[24,25] or water[154] permeability, lecithin content may be an important factor influencing the permeability of neutral molecules, such as glucose and glycerol.[29,90] It is clear, however, that effects of individual lipid classes on the permeability characteristics of thin lipid films does not

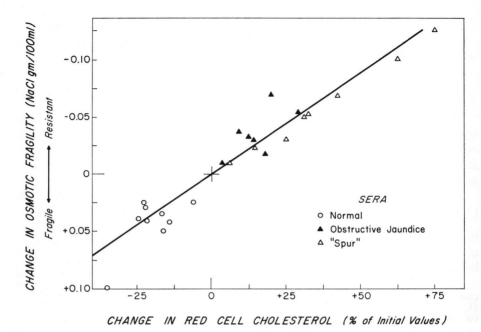

Fig. 2.—Influence of various sera on cholesterol content and osmotic fragility of normal red cells. Loss of cholesterol decreases membrane surface area causing red cells to be more osmotically fragile. Gains of cholesterol extend surface area and impart resistance to osmotic lysis.

predict what effect these lipids will have in a biological membrane.[154]

Experiments of nature have begun to provide insight into the functional and structural organization of the red cell membrane. Firm conclusions are not yet justified, but as the body of data grows the image brightens.

REFERENCES

1. Alaupovic, P., Seidel, D., McConathy, W. J., and Furman, R. H.: Identification of the protein moiety of an abnormal human plasma low-density lipoprotein in obstructive jaundice. FEBS Letters 4:113, 1969.

2. Ashworth, L. A. E., and Green, C.: The transfer of lipids between human α-lipoprotein and erythrocytes. Biochim. Biophys. Acta 84:182, 1964.

3. Bagdade, J. D., and Ways, P. O.: Erythrocyte membrane lipid composition in exogenous and endogenous hypertriglyceridemia. J. Lab. Clin. Med. 75:53, 1970.

4. Balint, J. A., Nyhan, W. L., Lietman, P., and Turner, D. A.: Lipid patterns in Niemann-Pick disease. J. Lab. Clin. Med. 58:548, 1961.

5. Balint, J. A., and Spitzer, H. L.: Lipid metabolism in Niemann-Pick's disease. Clin. Res. 10:31, 1962.

6. Balint, J. A., Spitzer, H. L., and Kyriakides, E. C.: Studies of red-cell stromal lipids in Tay-Sachs disease and other lipidoses. J. Clin. Invest. 42:1661, 1963.

7. Bar, R. S., Deamer, D. W., and Cornwell, D. G.: Surface area of human erythrocyte lipids: reinvestigation of experiments on plasma membrane. Science 153:1010, 1966.

8. Barrett, A. M.: A special form of erythrocyte possessing increased resistance to hypotonic salt. J. Path. Bact. 46:603, 1938.

9. Basford, J. M., Glover, J., and Green, C.: Exchange of cholesterol between human β-lipoproteins and erythrocytes. Biochim.

Biophys. Acta. 84:764, 1964.

10. Bassen, F. A., and Kornzweig, A. L.: Malformation of the erythrocytes in a case of atypical retinitis pigmentosa. Blood 5:381, 1950.

11. Berk, L.: Mechanism of red cell changes in non-hemolytic jaundice. Nature 155:269, 1945.

12. Bertles, J. F.: Sodium transport across the surface of membranes of red blood cells in hereditary spherocytosis. J. Clin. Invest. 36:816, 1957.

13. Brabec, V., Michalec, C., Palek, J., and Kout, M.: Red cell lipids in autoimmune hemolytic anemia. Blood 34:414, 1969.

14. Bradlow, B. A., Lee, J., and Rubenstein, R.: Erythrocyte phospholipids: quantitative thin-layer chromatography in paroxysmal nocturnal haemoglobinuria and hereditary spherocytosis. Brit. J. Haemat. 11:315, 1965.

14a. Brown, B. L.: Erythrocyte glycolipids in Huntington's chorea. Neurology 19:489, 1969.

15. Bruckdorfer, K. R., Graham, J. M., and Green, C.: The incorporation of steroid molecules into lecithin sols, β-lipoproteins and cellular membranes. European J. Biochem. 4:512, 1968.

16. —, and Green, C.: The exchange of unesterified cholesterol between human low-density lipoproteins and rat erythrocyte 'ghosts'. Biochem. J. 104:270, 1967.

17. Brun, G. C.: Cholesterol content of the red blood cells in man. Acta Med. Scand. Suppl. 99:1, 1939.

18. Burger, S. P., Fujii, T., and Hanahan, D. J.: Stability of the bovine erythrocyte membrane. Release of enzymes and lipid components. Biochemistry 7:3682, 1968.

19. Chan, P. C.: Reversible effect of sodium dodecyl sulfate on human erythrocyte membrane adenosine triphosphatase. Biochim. Biophys. Acta 135:53, 1967.

20. Cooper, R. A.: Loss of membrane components in the pathogenesis of antibody-induced spherocytosis. Clin. Res. 17:322, 1969.

20a. —: Unpublished observations.

21. —: Anemia with spur cells: A red cell defect acquired in serum and modified in the circulation. J. Clin. Invest. 48:1820, 1969.

21a. —, Admirand, W. H., Garcia, F., and Trey, C.: The role of lithocholic acid in the pathogenesis of spur red cells and hemolytic anemia. J. Clin. Invest. 47:18a, 1969.

22. —, and Jandl, J. H.: Bile salts and cholesterol in the pathogenesis of target cells in obstructive jaundice. J. Clin. Invest. 47:809, 1968.

23. —, and —: Physiologic and pathologic alterations of red cell lipids, membrane area, and shape. In Deutsch, E., Gerlach, E., and Moser, K., (Eds.): Metabolism and Membrane Permeability of Erythrocytes and Thrombocytes. Stuttgart, Georg Thieme, 1968, p. 376.

24. —, and —: The role of membrane lipids in the survival of red cells in hereditary spherocytosis. J. Clin. Invest. 48:736, 1969.

25. —, and —: The selective and conjoint loss of red cell lipids. J. Clin. Invest. 48:906, 1969.

26. —, and —: Red cell cholesterol content: a manifestation of the serum affinity for free cholesterol. Trans. Ass. Amer. Physicians 82:324, 1969.

27. Crowley, J., Ways, P., and Jones, J.: Human fetal erythrocyte and plasma lipids. J. Clin. Invest. 44:989, 1965.

28. Danielli, J. F., and Davson, H.: A contribution to the theory of permeability of thin films. J. Cell. Comp. Phys. 5:495, 1935.

29. De Gier, J., and Van Deenen, L. L. M.: Some lipid characteristics of red cell membranes of various animal species. Biochim. Biophys. Acta 49:286, 1961.

30. —, —, Verloop, M. C., and Van Gastel, C.: Phospholipid and fatty acid characteristics of erythrocytes in some cases of anaemia. Brit. J. Haemat. 10:246, 1964.

31. Dervichian, D. G., and Macheboeuf, M.: Sur l'existence d'une couche monomoléculaire de substances lipoïdiques á la surface des globules rouges du sang. C. R. Sean. Acad. Scien. 206:1511, 1938.

32. De Venuto, F.: Interaction of progesterone and aldosterone with red blood cells of the rat. Proc. Soc. Exp. Biol. Med. 124:478, 1967.

33. —: Osmotic fragility and spontaneous lysis of human red cells preserved with addition of progesterone. Proc. Soc. Exp. Biol. Med. 128:997, 1968.

34. Dodge, J. T., Cohen, G., Kayden, H. J., and Phillips, G. B.: Peroxidative hemolysis of red blood cells from patients with abetalipoproteinemia (acanthocytosis). J. Clin. Invest. 46:357, 1967.

35. —, Mitchell, C., and Hanahan, D. J.: The preparation and chemical characteristics of hemoglobin-free ghosts of human erythrocytes. Arch. Biochem. Biophys. 180:119, 1963.

36. —, and Phillips, G. B.: Autoxidation as a cause of altered lipid distribution in extracts from human red cells. J. Lipid Res. 7:387, 1966.

37. —, and —: Composition of phospholipids and of phospholipid fatty acids and aldehydes in human red cells. J. Lipid Res. 8:667, 1967.

38. Donabedian, R. K., and Karmen, A.: Fatty acid transport and incorporation into human erythrocytes in vitro. J. Clin. Invest. 46:1017, 1967.

39. Engelman, D. M.: Surface area per lipid molecule in the intact membrane of the human red cell. Nature 223:1279, 1969.

40. Erickson, B. N., Williams, H. H., Bernstein, I. A., Jones, R. L., and Macy, I. C.: The lipid distribution of posthemolytic residue or stroma of erythrocytes. J. Biol. Chem. 122:515, 1958.

41. Farnsworth, P., Danon, D., and Gellhorn, A.: The effect of fat-free diet on the rat erythrocyte membrane. Brit. J. Haemat. 11:200, 1965.

42. Farquhar, J. W.: Human erythrocyte phosphoglycerides: I. Quantification of plasmalogens, fatty acids and fatty aldehydes. Biochim. Biophys. Acta 60:80, 1962.

43. —, and Ahrens, H., Jr.: Effects of dietary fats on human erythrocyte fatty acid patterns. J. Clin. Invest. 42:675, 1963.

44. Fels, G., Kanabrocki, E., and Kaplan, E.: Plasma and red cell cholesterol. Clin. Chem. 7:16, 1939.

45. Fenster, L. J., and Copenhaver, J. H., Jr.: Phosphatidyl serine requirement of (Na+—K+)—activated adenosine triphosphatase from rat kidney and brain. Biochim. Biophys. Acta 137:406, 1967.

46. Fleischer, S., Fleischer, B., and Rouser, G.: Quoted in Rouser, G., Nelson, G. J., Fleischer, S., and Simon, G.: Lipid composition of animal cell membranes, organelles, and organs. In Chapman, D., (Ed.): Biological Membranes, Physical Fact and Function. New York, Academic Press, 1968, p. 5.

47. —, Rouser, G., Fleischer, B., Casu, A., and Kritchevsky, G.: Lipid composition of mitochondria from bovine heart, liver, and kidney. J. Lipid Res. 8:170, 1967.

48. Folch, J., Lees, M., and Stanley, G. H. S.: A simple method for the isolation and purification of total lipids from animal tissues. J. Biol. Chem. 226:497, 1957.

49. Formijne, P., Poulie, N. J., and Rodbard, J. A.: Determination of phospholipid fractions in the human erythrocyte. Clin. Chim. Acta 2:25, 1957.

50. Fredrickson, D. S.: Familial high-density lipoprotein deficiency: Tangier disease. In Stanbury, J. B., Wyngaarden, J. B., and Fredrickson, D. S., (Eds.): The Metabolic Basis of Inherited Disease (2nd ed.) New York, McGraw-Hill, 1966, p. 486.

51. —, Altrocchi, P. H., Avioli, L. V., Goodman, D. S., and Goodman, H. C.: Tangier disease. Ann. Intern. Med. 55:1016, 1961.

52. Frezal, J., Rey, J., Polonovski, J., Levy, G., and Lamy, M.: L'absence congénitale de β-lipoprotéines: ètude de l'absorption des graisses après exsanguino-transfusion mesure de la demi-vil des β-lipoprotéines injectées. Rev. Franc. Clin. Biol. 6:677, 1961.

53. Geill, T.: Studies over ikterus IV. Undersogelsen over erythrocyternes osmotik resistens ved liver-og galdenejslidelser specielt ved de akute hepatitis. Hospitalstidende 74:433, 1931.

54. Gjone, E., and Norum, K. R.: Familial serum cholesterol ester deficiency: clinical study of a patient with a new syndrome. Acta Med. Scand. 183:107, 1968.

55: —, Torsvik, H., and Norum, K. R.: Familial plasma cholesterol ester deficiency: a study of the erythrocytes. Scand. J. Clin. Lab. Invest. 21:327, 1968.

56. Glomset, J. A.: The plasma lecithin: cholesterol acyltransferase reaction, J. Lipid Res. 9:155, 1968.

57. Gorter, E., and Grendel, F.: On biomolecular layers of lipoids on the chromocytes of the blood. J. Exp. Med. 41:439, 1925.

58. Gould, R. G., Le Roy, G. V., Okita, G. T., Kabara, J. J., Keegan, P., and Bergenstal, D. M.: The use of C14-labeled acetate to study cholesterol metabolism in man. J. Lab. Clin. Med. 46:372, 1955.

59. Graham, J. M., and Green, C.: The binding of hormones and related compounds by normal and cholesterol-depleted plasma membranes of rat liver. Biochem. Pharmacol. 18:493, 1969.

60. Grossman, C. M., Horky, J., and

Kohn, R.: In vitro incorporation of ³²P-orthophosphate into phosphatidyl ethanolamine and other phosphatides by mature human erythrocyte ghosts. Arch. Biochem. 117:18, 1966.

61. Hagerman, J. S., and Gould, R. G.: The in vitro interchange of cholesterol between plasma and red cells. Proc. Soc. Exp. Biol. Med. 78:329, 1951.

62. Hakomori, S. I., and Strycharz, G. D.: Investigations on cellular blood group substances. I. Isolation and chemical composition of blood group ABH and Le^b isoantigens of sphingoglycolipid nature. Biochemistry 7:1279, 1968.

63. Hanahan, D. J.: Characterization of the erythrocyte membrane. In Jamieson, G. A., and Greenwalt, T. J., (Eds.): Red Cell Membrane Structure and Function. Philadelphia, J. B. Lippincott, 1969, pp. 83–92.

64. Haradin, A. R., Weed, R. I., and Reed, C. F.: Changes in physical properties of stored erythrocytes: relationship to survival in vivo. Transfusion 9:229, 1969.

65. Harris, I. M., Prankerd, T. A. J., and Westerman, M. P.: Abnormality of phospholipids in red cells of patients with paroxysmal nocturnal haemoglobinuria. Brit. Med. J. 2:1276, 1957.

66. Harris, J. W., and Schilling, R. F.: Increased resistance to osmotic lysis as an acquired change in the erythrocytes of patients with hepatogenous jaundice or biliary obstruction. J. Clin. Invest. 26:820, 1950.

67. Hestorff, R., Ways, P., and Palmer, S.: The pathophysiology of "cholesterol anemia" in rabbits. J. Clin. Invest. 44:1059, 1965.

68. Hoffman, H. N., and Fredrickson, D. S.: Tangier disease (familial high density lipoprotein deficiency): clinical and genetic features in two adults. Amer. J. Med. 39:582, 1965.

69. Hokin, L. E., and Hokin, M. R.: The incorporation of ³²P from triphosphate into polyphosphinoinositides [γ-³²P] adenosine and phosphatidic acid in erythrocyte membranes. Biochim. Biophys. Acta 84:563, 1964.

70. Hooghwinkel, G. J. M., Borri, P. F., and Bruyn, G. W.: Biochemical studies in Huntington's chorea. V. Erythrocyte and plasma glycolipids and fatty acid composition of erythrocyte gangliosides. Neurology 16:934, 1968.

71. Ishii, J., and Nakao, K.: Analysis of the human erythrocyte lipid in various functional conditions of the thyroid. Tohoku J. Exp. Med. 94:169, 1968.

72. Jacob, H. S.: Membrane lipid depletion in hyperpermeable red blood cells: its role in the genesis of spherocytes in hereditary spherocytosis. J. Clin. Invest. 46:2083, 1967.

73. —, and Jandl, J. H.: Increased cell membrane permeability in the pathogenesis of hereditary spherocytosis. J. Clin. Invest. 43:1704, 1964.

74. —, and Karnovsky, M. L.: Concomitant alterations of sodium flux and membrane phospholipid metabolism in red blood cells: studies in hereditary spherocytosis. J. Clin. Invest. 46:173, 1967.

75. —, and Lux, S. E. IV,: Degradation of membrane phospholipids and thiols in peroxide hemolysis: studies in vitamin E deficiency. Blood 32:549, 1968.

76. Jaffé, E. R., and Gottfried, E. L.: Hereditary nonspherocytic hemolytic disease associated with an altered phospholipid composition of the erythrocytes. J. Clin. Invest. 47:1375, 1968.

77. Jandl, J. H.: The anemia of liver disease: observations on its mechanisms. J. Clin. Invest. 34:390, 1955.

78. Jones, J. W., and Ways, P.: Abnormalities of high density lipoproteins in abetalipoproteinemia. J. Clin. Invest. 46:1151, 1967.

79. Karlsson, K., Samuelsson, B. E., and Steen, G. O.: Structure and function of sphingolipids: 2. Differences in sphingolipid concentration, especially concerning sulfatides, between some regions of bovine kidney. Acta. Chem. Scand. 22:2723, 1968.

80. Kates, M., Allison, A. C., and James, A. T.: Phosphatides of human blood cells and their role in spherocytosis. Biochim. Biophys. Acta 48:571, 1961.

81. Kayden, H. J., and Silber, R.: The role of vitamin E deficiency in the abnormal autohemolysis of acanthocytosis. Trans. Ass. Amer. Physicians 78:334, 1965.

82. Kessler, J. I., Kniffer, J. C., and Janowitz, H. D.: Lipoprotein lipase inhibition in hyperlipemia of acute alcoholic pancreatitis. New Eng. J. Med. 269:943, 1963.

83. Kimber, C., Deller, D. J., Ibbotson, R. N., and Lander, H.: The mechanism of anaemia in chronic liver disease. Quart. J. Med. 34:33, 1965.

84. Korn, E. D.: Structure of biological membranes: The unit membrane theory is reevaluated in light of the data now available. Science 153:1491, 1966.

85. Krivit, W., and Kern, L.: Intravascular membrane dehiscence of red cell: model survival curve using glycolipid labelled with C^{14} glucose in rabbits. Blood 34:858, 1969.

86. Kurland, G. S., Lucas, J. C., and Friedberg, A. S.: The metabolism of intravenously infused C^{14}-labeled cholesterol in euthyroidism and myxedema. J. Lab. Clin. Med. 57:574, 1961.

87. LaCelle, P. L., and Weed, R. I.: Reversibility of abnormal deformability and permeability of the hereditary spherocyte. Blood 34:858, 1969.

88. Langley, G. R., and Axell, M.: Changes in erythrocyte membrane and autohaemolysis during in vitro incubation. Brit. J. Haemat. 14:593, 1968.

89. —, and Felderhof, C. H.: Atypical autohemolysis in hereditary spherocytosis as a reflection of two cell populations: Relationship of cell lipids to conditioning by the slpeen. Blood 32:569, 1968.

90. Lares, P. C.: Permeability of and utilization of glucose in mammalian erythrocytes. J. Cell. Comp. Physiol. 51:273, 1958.

91. Levy, R. I., Fredrickson, D. S., and Laster, L.: The lipoproteins and lipid transport in abetalipoproteinemia. J. Clin. Invest. 45:531, 1966.

92. London, I. M., and Schwartz, H.: Erythrocyte metabolism. The metabolic behavior of the cholesterol of human erythrocytes. J. Clin. Invest. 32:1248, 1953.

93. Losowsky, M. S., Jones, D. P., Davidson, C. S., and Lieber, C. S.: Studies of alcoholic hypertriglyceridemia and its mechanism. Amer. J. Med. 35:794, 1963.

94. Luzzati, V., Gulik-Krzywicki, T., and Tardieu, A.: Polymorphism of lecithins. Nature 218:1031, 1968.

95. Marcus, D. M., and Cass, L. E.: Glycosphingolipids with Lewis blood group activity: uptake by human erythrocytes. Science 164:553, 1969.

96. Marks, P. A., Gellhorn, A., and Kidson, C.: Lipid synthesis in human leukocytes, platelets, and erythrocytes. J. Biol. Chem. 235:2579, 1960.

97. Marvin, H. N.: Some metabolic and nutritional factors affecting survival time of erythrocytes. Amer. J. Clin. Nutr. 12:88, 1963.

98. McBride, J. A., and Jacob, H. S.: Cholesterol loading of acanthocytic red cell membranes causing hemolytic anemia in experimental and genetic abetalipoproteinemia. J. Clin. Invest. 47:67a, 1968.

99. MeLeod, M. E., and Bressler, R.: Some aspects of phospholipid metabolism in the red cell. Biochim. Biophys. Acta 144:391, 1967.

100. Miller, E. B., Singer, K., and Dameshek, W.: Experimental production of target cells by splenectomy and interference with splenic circulation. Proc. Soc. Exp. Biol. Med. 49:42, 1942.

101. Mulder, E., and Van Deenen, L. L. M.: Metabolism of red cell lipids: I. Incorporation in vitro of fatty acids into phospholipids from mature erythrocytes. Biochim. Biophys. Acta 106:106, 1965.

102. —, Van den Berg, J. W. O., and Van Deenen, L. L. M.: Metabolism of red-cell lipids. II. Conversions of lysophosphoglycerides. Biochim. Biophys. Acta 106:118, 1965.

103. Murphy, J. R.: Erythrocyte metabolism. III. Relationship of energy metabolism and serum factors to the osmotic fragility following incubation. J. Lab. Clin. Med. 60:86, 1962.

104. —: Erythrocyte metabolism. IV. Equilibration of cholesterol—4—C^{14} between erythrocytes and variously treated sera. J. Lab. Clin. Med. 60:571, 1962.

105. —: Erythrocyte metabolism. VI. Cell shape and the location of cholesterol in the erythrocyte membrane. J. Lab. Clin. Med. 65:756, 1965.

106. Nakao, K., Kurashina, S., and Nakao, M.: Adenosinetriphosphatase activity of erythrocyte membrane in hereditary spherocytosis. Life Sci. 6:595, 1967.

107. Neerhout, R. C.: Erythrocyte stromal lipids in hyperlipemic states. J. Lab. Clin. Med. 71:445, 1968.

108. —: Abnormalities of erythrocyte stromal lipids in hepatic disease: erythrocyte stromal lipids in hyperlipemic states. J. Lab. Clin. Med. 71:438, 1968.

109. —: Erythrocyte lipids in the neonate. Pediat. Res. 2:172, 1968.

110. —: Reversibility of the erythrocyte lipid abnormalities in hepatic disease. J. Pediat. 73:364, 1968.

111. Nelson, G. J.: Lipid composition of

erythrocytes in various mammalian species. Biochim. Biophys. Acta 144:221, 1967.

112. —: Composition of neutral lipids from erythrocytes of common mammals. J. Lipid Res. 8:374, 1967.

113. —: Studies on the lipids of sheep red blood cells I. Lipid composition in low and high potassium red cells. Lipids 2:64, 1967.

114. —: Studies on the lipids of sheep red blood cells: III. The fatty acid composition of phospholipids in HK and LK cells. Lipids 4:350, 1969.

115. Nervi, A. M., and Brenner, R. R.: Rate of linoleic and arachidonic acid incorporation into liver, heart and red cells of essential fatty acid deficient rats and its effect on eicosatrienoic acid depletion. Acta Physiol. Lat. Amer. 15:308, 1965.

116. Nestle, P. J., and Couzens, E. A.: Turnover of individual cholesterol esters in human liver and plasma. J. Clin. Invest. 45:1234, 1966.

117. —, and Monger, E. A.: Turnover of plasma esterified cholesterol in normocholesterolemic and hypercholesterolemic subjects and its relation to body build. J. Clin. Invest. 46:967, 1967.

118. Norum, K. R.: Abnormalities in plasma lipoproteins in patients with familial plasma lecithin:cholesterol acyltransferase (LCAT) deficiency. Biophys. J. 9:A-149, 1969.

119. —, and Gjone, E.: Familial plasma lecithin: cholesterol acyltransferase deficiency: biochemical study of a new inborn error of metabolism. Scand. J. Clin. Lab. Invest. 20:231, 1967.

120. —, and —: Familial serum-cholesterol esterification failure. A new inborn error of metabolism. Biochim. Biophys. Acta 144:698, 1967.

121. —, and —: The influence of plasma from patients with familial plasma lecithin: cholesterol acyltransferase deficiency on the lipid pattern of erythrocytes. Scand. J. Clin. Lab. Invest. 22:94, 1968.

122. Nye, W. H. R., and Marinetti, G. V.: Alterations in erythrocyte phospholipids produced by environmental change. Proc. Soc. Exp. Biol. Med. 125:1220, 1967.

123. Okey, R.: Cholesterol injury in the guinea pig. J. Biol. Chem. 156:179, 1944.

124. —, and Greaves, V. D.: Anemia caused by feeding cholesterol to guinea pigs. J. Biol. Chem. 129:111, 1939.

125. Oliveira, M. M., and Vaughan, M.: Incorporation of fatty acids into phospholipids of erythrocyte membranes. J. Lipid Res. 5:156, 1964.

126. Oski, F. A., and Barness, L. A.: Vitamin E deficiency: a previously unrecognized cause of hemolytic anemia in the premature infant. J. Pediat. 70:211, 1967.

127. —, Naiman, J. L., Blum, S. F., Zarkowsky, H. S., Whann, J., Shohet, S. B., Green, A., and Nathan, D. G.: Congenital hemolytic anemia with high-sodium, low-potassium red cells. New Eng. J. Med. 280:909, 1969.

128. Ostwald, R., and Shannon, A.: Composition of tissue lipids and anaemia of guinea pigs in response to dietary cholesterol. Biochem. J. 91:146, 1964.

129. Pak, C. Y. C., and Gershfeld, N. L.: Steroid hormones and monolayers, Nature 214:888, 1967.

130. Parpart, A. K., and Dziemian, A. J.: The chemical composition of the red cell membrane. Cold Spring Harbor Symposia on Quant. Biol. 8:17, 1940.

131. Phillips, G. B.: Quantitative chromatographic analysis of plasma and red cell lipids in patients with acanthocytosis. J. Lab. Clin. Med. 59:357, 1962.

132. —, Dodge, J. T., and Rockmore, C. S.: Analysis of fatty acids of human red cells without lipid extraction. J. Lipid Res. 9:285, 1968.

133. —, and Roome, N. S.: Quantitative chromatographic analysis of the phospholipids of abnormal human red cells. Proc. Soc. Exp. Biol. Med. 109:360, 1962.

134. Pinter, G. G., and Bailey, R. E.: Anemia of rabbits fed a cholesterol-containing diet. Amer. J. Physiol. 200:292, 1961.

135. Pittman, J. G., and Martin, D. B.: Fatty acid biosynthesis in human erythrocytes: evidence in mature erythrocytes for an incomplete long chain fatty acid synthesizing system. J. Clin. Invest. 45:165, 1966.

136. Porte, D., Jr., and Havel, R. J.: The use of cholesterol-4-C^{14} labeled lipoproteins as a tracer for plasma cholesterol in the dog. J. Lipid Res. 2:357, 1961.

137. Rand, R. P., and Burton, A. C.: Mechanical properties of the red cell membrane. I. Membrane stiffness and intracellular pressure. Biophysical J. 4:115, 1964.

138. Reed, C. F.: Incorporation of orthophosphate-^{32}P into erythrocyte phospholipids in normal subjects and in patients with

hereditary spherocytosis. J. Clin. Invest. 47:2630, 1968.

139. —: Phosholipid exchange between plasma and erythrocytes in man and the dog. J. Clin. Invest. 47:749, 1968.

140. —, Eden, E. G., and Swisher, S. N.: Studies of the lipids of normal and stored human red cells. Clin. Res. 6:186, 1958.

141. —, and Swisher, S. N.: Erythrocyte lipid loss in hereditary spherocytosis. J. Clin. Invest. 45:777, 1966.

142. —, —, Marinetti, G. V., and Eden, E. G.: Studies of the lipids of the erythrocyte I. Quantitative analysis of the lipids of normal human red blood cells. J. Lab. Clin. Med. 56:281, 1960.

143. Robertson, J. D.: Unit membranes: A review with recent new studies of experimental alterations and a new subunit structure in synaptic membranes. In Locke, M., (Ed.): Cellular Membranes in Development. New York, Academic Press, 1964, p. 1.

144. Robertson, A. F., and Lands, W. E. M.: Metabolism of phospholipids in normal and spherocytic human erythrocytes. J. Lipid Res. 5:88, 1964.

145. Roelofsen, B., Baadenhuysen, H., and van Deenen, L. L. M.: Effects of organic solvents on the adenosine triphosphatase activity of erythrocyte ghosts. Nature 212:1379, 1966.

146. Rose, H. G., and Oklander, M.: Improved procedure for the extraction of lipids from human erythrocytes. J. Lipid Res. 6:428, 1965.

147. Rowe, C. E.: The biosynthesis of phospholipids by human blood cells. Biochem. J. 73:438, 1959.

148. —, Allison, A. C., and Lovelock, J. E.: Synthesis of lipids by different human blood cell types. Biochim. Biophys. Acta 41:310, 1960.

149. Russ, E. M., Raymunt, J., and Barr, D. P.: Lipoproteins in primary biliary cirrhosis. J. Clin. Invest. 35:133, 1956.

150. Sakagami, T., Minari, D., and Orii, T.: Interaction of individual phospholipids between rat plasma and erythrocytes in vitro. Biochim. Biophys. Acta 98:356, 1965.

150a. Salen, G., Ahrens, E. H., Jr., and Grundy, S.: The metabolism of β-sitosterol in man. J. Clin. Invest. 49:952, 1970.

151. Salt, H. B., Wolff, O. H., Llyd, J. K., Fosbrooke, A. S., Cameron, A. H., and Hubble, D. V.: On having no beta-lipoprotein. A syndrome comprising a-beta-lipoproteinemia, acanthocytosis, and steatorrhoea. Lancet 2:325, 1960.

152. Schwartz, J. F., Rowland, L. P., Eden, H., Marks, P. A., Osserman, E. F., Herschberg, E., and Anderson, H.: Bassen-Kornzweig syndrome:deficiency of serum beta-lipoprotein. A neuromuscular disorder resembling Friedreich's ataxia, associated with steatorrhea, acanthocytosis, retinitis pigmentosa, and a disorder of lipid metabolism. Arch. Neurol. 8:438, 1963.

153. Seidel, D., Alaupovic, P., and Furman, R. H.: A lipoprotein characterizing obstructive jaundice. I. Method for quantitative separation and identification of lipoproteins in jaundiced subjects. J. Clin. Invest. 48:1211, 1969.

153a. Sears, D. A., Reed, C. F., and Weed, R. I.: Unpublished observations.

154. Sha'afi, R. I., Gary-Bobo, C., and Solomon, A. K.: Cholesterol effect on hydraulic conductivity of red cell membranes. Biochim. Biophys. Acta 173:141, 1969.

155. Shacklady, M. M., Djardjouras, E. M., and Lloyd, J. K.: Red-cell lipids in familial alphalipoprotein deficiency (Tangier disease). Lancet 2:151, 1968.

156. Shohet, S. B.: The release of phospholipid fatty acid from human erythrocytes. J. Clin. Invest. (in press).

157. —, Livermore, B. M., and Jaffé, E. R.: The mechanism of phosphatidyl choline (PC) accumulation in the red blood cells of patients with familial hemolytic anemia and altered erythrocyte lipids. Blood 34:859, 1969.

158. —, Nathan, D. G., and Karnovsky, M. L.: Stages in the incorporation of fatty acids into red blood cells. J. Clin. Invest. 47:1096, 1968.

159. Silber, R., Amorosi, E., Lhowe, J., and Kayden, H. J.: Spur-shaped erythrocytes in Laennec's cirrhosis. New Eng. J. Med. 275:639, 1966.

160. —, Winter R., and Kayden, H. J.: Tocopherol transport in the rat erythrocyte. J. Clin. Invest. 48:2089, 1969.

161. Silver, M. M., McMillan, G. C., and Silver, M. D.: Haemolytic anaemia in cholesterol-fed rabbits. Brit. J. Haemat. 10:271, 1964.

162. Simon, E. R., and Ways, P.: Incubation hemolysis and red cell metabolism in acanthocytosis. J. Clin. Invest. 43:1311, 1964.

163. Singer, K., Fisher, B., and Perlstein,

M. A.: Acanthrocytosis, a genetic erythrocytic malformation. Blood 7:577, 1952.

164. —, Miller, E. B., and Dameshek, W.: Hematological changes following splenectomy in man, with particular reference to target cells, hemolytic index and lysolecithin. Amer. J. Med. Sci. 202:171, 1941.

165. Skou, J. C.: Enzymatic basis for active transport of Na+ and K+ across cell membranes. Physiol. Rev. 45:596, 1965.

166. Smith, J. A., Lonergan, E. T., and Sterling, K.: Spur-cell anemia: hemolytic anemia with red cells resembling acanthocytes in alcoholic cirrhosis. New Eng. J. Med. 271:396, 1964.

167. Sperry, W. M.: Cholesterol esterase in blood. J. Biol. Chem. 111:467, 1935.

168. Spritz, N.: Effect of fatty acid saturation on the distribution of the cholesterol moiety of very low density lipoproteins. J. Clin. Invest. 44:339, 1965.

169. Stoeckenius, W., and Engelman, D. M.: Current models for the structure of biological membranes. J. Cell Biol. 42:613, 1969.

170. Sweeley, C. C., and Dawson, G.: Lipids of the erythrocyte. In Jamieson, G. A., and Greenwalt, T. J., (Eds.): Red Cell Membrane Structure and Function. Philadelphia, J. B. Lippincott, 1969, p. 172.

171. Switzer, S.: Plasma lipoproteins in liver disease: I. Immunologically distinct low-density lipoproteins in patients with biliary obstruction. J. Clin. Invest. 46:1855, 1967.

172. —: Plasma lipoproteins in the differential diagnosis of liver disease. Gastroenterology 53:790, 1967.

173. —, and Eder, H.: Interconversion of acanthocytes and normal erythrocytes with detergents. J. Clin. Invest. 41:1404, 1962.

174. Szeinberg, A., Zaidman, J., and Clejan, L.: Investigation of the lipid content of normal and glucose-6-phosphate dehydrogenase deficient red cells. Biochim. Biophys. Acta 98:598, 1965.

175. Tarlov, A. R., and Mulder, E.: Phospholipid metabolism in rat erythrocytes: Quantitative studies of lecithin biosynthesis. Blood 30:853, 1967.

176. Torsvik, H.: Presence of α_1-lipoprotein in patients with familial plasma lecithin:cholesterol acyltransferase deficiency. Scand. J. Clin. Lab. Invest. 24:187, 1969.

177. —, Gjone, E., and Norum, K. R.: Familial plasma cholesterol ester deficiency. Acta Med. Scand. 183:387, 1968.

178. Tsai, C., Lee, J-S, and Wu, C. H.: The role of splenic action in altering erythrocyte fragility. Chinese J. Physiol. 15:165, 1940.

179. Turner, K. B., McCormack, G. H., and Richards, A.: The cholesterol-esterifying enzyme of human serum. I. In liver disease. J. Clin. Invest. 32:801, 1953.

180. Van Gastel, C., Van den Berg, B., De Gier, J., and Van Deenen, L. L. M.: Some lipid characteristics of normal red blood cells of different age. Brit. J. Haemat. 11:193, 1965.

181. Waku, K., and Lands, W. E. M.: Control of lecithin biosynthesis in erythrocyte membranes. J. Lipid Res. 9:12, 1968.

182. Walker, B. L.: Recovery of rat tissue lipids from essential fatty acid deficiency: plasma, erythrocytes and liver. J. Nutrition 92:23, 1967.

183. Watson, W. C.: The morphology and lipid composition of the erythrocyte in normal and essential-fatty-acid deficient rats. Brit. J. Haemat. 9:32, 1963.

184. Ways, P.: An acquired reversible abnormality of erythrocyte lipids associated with liver disease and hemolytic anemia. J. Clin. Invest. 46:1129, 1967.

185. —: Degradation of glycerophosphatides during storage of saline-washed, saline-suspended red cells at −20°C. J. Lipid Res. 8:518, 1967.

186. —, and Hanahan, D. J.: Characterization and quantification of red cell lipids in normal man. J. Lipid Res. 5:318, 1964.

187. —, Reed, C. F., and Hanahan, D. J.: Red-cell and plasma lipids in acanthocytosis. J. Clin. Invest. 42:1248, 1963.

188. Weed, R. I., Bowdler, A. J., and Reed, C. F.: Metabolic dependence of erythrocyte membrane structure. Clin. Res. 13:284, 1965.

189. —, LaCelle, P. L., and Merrill, E. W.: Metabolic dependence of red cell deformability. J. Clin. Invest. 48:795, 1969.

190. Westerman, M. P., Balcerzak, S. P., and Heinle, E. W., Jr.: Red cell lipids in Zieve's syndrome: their relation to hemolysis and to red cell osmotic fragility. J. Lab. Clin. Med. 72:663, 1968.

191. —, and Jensen, W. N.: In vitro incorporation of radiophosphorus into the phosphatides of normal human blood cells. Proc. Soc. Exp. Biol. Med. 118:315, 1965.

192. —, Pierce, L. E., and Jensen, W. N.: Erythrocyte lipids: a comparison of normal young and normal old populations. J. Lab. Clin. Med. 62:394, 1963.

193. —, —, and —: Erythrocyte and plasma lipids in sickle cell anemia. Blood 23:200, 1964.

193a. Wiley, J. S.: Dominant inheritance of the sodium leak in hereditary spherocytic red cells and a comparison of sodium leak with red cell survival. Aust. Ann. Med. 17:177, 1968.

194. —: Inheritance of an increased sodium pump in human red cells. Nature, 221:1222, 1969.

195. Willmer, E. N.: Steroids and cell surfaces. Biol. Rev. 36:368, 1961.

196. Wilson, D. E., Schreibman, P. H., Brewster, A. C., and Arky, R. A.: The enhancement of alimentary lipemia by ethanol in man. J. Lab. Clin. Med. 75:264, 1970.

197. Windmueller, H. G., and Levy, R. I.: Total inhibition of hepatic β-lipoprotein production in the rat by orotic acid. J. Biol. Chem. 242:2246, 1967.

198. Winzler, R. J.: A glycoprotein in human erythrocyte membranes. In Jamieson, G. A., and Greenwalt, T. J., (Eds.): Red Cell Membrane Structure and Function. Philadelphia, J. B. Lippincott, 1969, p. 157.

199. Wittels, B., and Hochstein, P.: The effect of primaquine on lecithin metabolism in human erythrocytes. Biochim. Biophys. Acta 125:594, 1966.

199a. Yamakawa, T.: Glycolipids of mammalian red blood cells. In Schütte, C. E. (Ed.): Lipoide:Glycolipids of Mammalian Red Blood Cells. New York, Springer, 1966, p. 87.

200. Zarkowsky, H., Oski, F., Sha'afi, R., Shohet, S. B., and Nathan, D. G.: Congenital hemolytic anemia with high sodium, low potassium red cells. New Eng. J. Med. 278:573, 1968.

201. Zieve, L.: Jaundice, hyperlipemia and hemolytic anemia: a heretofore unrecognized syndrome associated with alcoholic fatty liver and cirrhosis. Ann. Intern. Med. 48:471, 1958.

202. —: Hemolytic anemia in liver disease. Medicine 45:497, 1966.

Effects of Immune Reactions on the Red Cell Membrane

By WENDELL F. ROSSE AND PETER K. LAUF

T HE RED CELL has been used to detect immunological reactions almost since the beginning of the study of such reactions. The reasons for this popularity are several: Red cells are easily "purified" and do not naturally adhere to one another; they can be used readily in tests employing agglutination. The nature of the coat of the membrane is such that other materials can be passively adsorbed to the surface of the red cell, thus permitting the study of immunological reactions involving antigens and antibody not native to the red cell itself. Further, cellular disruption is clearly marked by the release of hemoglobin. Red cells, therefore, have been used almost exclusively in the study of the immunological reactions resulting in cytolysis. However, throughout many of these studies the red cell has been regarded as a passive part of the system and the main attention has been focused upon the antigen-antibody reactions and the reactions of complement. Only recently have efforts been made to determine the effect of immunological reactions on the red cell membrane itself.

Immunological reactions related to the red cell surface may be divided into those involving antibody alone and those involving the components of serum complement.

The presence of antigens upon the red cell surface permits the fixation of antibody molecules. In certain circumstances, the binding of antibody triggers the activation of the complement system. By sequential enzymatic reactions, its components are bound to the red cell surface and ultimately lyse the red cell. These immunological reactions are enhanced by the organization at the surface, but at the same time, may alter the surface upon which they occur. It is the nature of these alterations with which we will be most concerned.

THE EFFECT OF ANTIBODIES

Since the first observation of blood group antigens on human erythrocytes by Karl Landsteiner in 1900,[41] a large number of homo- and heterologous antibodies directed against erythrocyte surface antigens have been discovered, often as the result of problems in blood group compatibility seen in blood transfusions, the hemolytic disorders of the newborn and autoimmune hemolytic diseases. In the last decade, a considerable body of knowledge has been accumulated concerning the structure and biological properties of blood group antibodies[8,38,61] and it is well established that antigenic groupings

From the Departments of Medicine and Microbiology, and Physiology and Pharmacology, Duke University Medical Center, Durham, N.C.

WENDELL F. ROSSE, M.D.: *Associate Professor of Medicine and Immunology, Duke University Medical Center, Durham, N.C.* PETER K. LAUF, M.D.: *Assistant Professor of Physiology and Immunology, Duke University Medical Center, Durham, N.C.*

at the outer surface of the erythrocyte are their prime target. Information on the nature of some of these antigenic determinants has been obtained by immunochemical studies of intact and enzymatically treated red cells[62,103,111] and their extracted membrane components.[39,113,104] In some instances, studies on mucopolysaccharides of certain biological fluids specifically inhibiting anti-red cell antibodies indicate that similar carbohydrates are present as antigens on the red cell surface.[63] This observation has been confirmed, in part by identification of these carbohydrates in the glycolipid fraction of the red cell membrane.[29,40,53]

The biological role of any of the blood group antigens is not yet established, but in all probability they may have specific functions unrelated to their immunogenic properties. The only role presently ascribed to surface antigens is that of the regulation of cell to cell interaction. The erythrocyte surface behaves like a polyanion due to the presence of terminally bound siliac acid (neuraminic acid), either in N-acetyl or N-glycolyl form.[20,92] At least some of these groups are contributed by the M and N antigens since n-acetyl-neuraminic acid is crucial to their antigenic structure.[104,105,113,114] The negative net surface charge imparted by the ionized carboxyl groups of neuraminic acid is responsible for the migration of the cells toward the anode in a direct current field.[25,26,86] The electrostatic repulsive forces between the red cells are due, in some degree, to the negative surface charge but to a greater degree to the so-called zeta-potential. The zeta- or electrokinetic potential arises as a result of interaction of the cations of the suspending medium with the charged surface of the red cell. This "ionic cloud" decreases as the distance from the densest zone of the charged double layer increases.[73]

Sensitization of human red cells with antibodies results in about a 30 per cent decrease in negative charge with a resultant slight decrease in repulsive forces. Agglutination occurs when the distance imposed by the electrostatic repulsive forces can be bridged by bivalent antibodies. Thus, for the 19S antibodies, the reduction in charge alone is sufficient to allow agglutination of the cells in saline suspension. Depending on the antigen site density,[34] agglutination by 7S antibodies may require further lowering of the electrostatic repulsive force by reduction of the zeta-potential. If the antigen site density is sufficiently high, IgG antibodies can produce agglutination in saline but if there are less than 20,000 sites per cell, saline agglutination will not occur unless the dielectric constant of the medium[73] is raised with anisotropic molecules in solution. Alteration of the surface charge of the red cell is a general action of antibodies on the red cell membrane resulting in changes of the cell-to-cell interaction.

The structural alteration consequent upon reaction with antibody has been studied with both light and electron microscopy, but it was not until the development of the stereoscan electron microscope that the events could be clearly seen. When cells were reacted with anti-A antibody, the surface quickly became roughened and finger-like processes were seen to "extrude" from the surface.[87] These processes appeared to be sticky and, when they occurred in neighboring cells, they appeared to bring the cells together.

Table 1.—Antibodies and Changes in Membranes Ascribed to Antibody-Antigen-Reactions

Antibody	Antigenic Determinants	Membrane Changes Observed	Reference
Human anti-A	Carbohydrates. Specificity: N-acetyl-galactosamine	Decrease in ATP content, Increase ATPase activity	70
		Ultrastructural changes	87
Human anti-I	Probably carbohydrate	Reversible ultrastructural changes	88
Human anti-D	Unknown, antigen on D (Rh$_0$) positive red cells and membranes	Decrease in glycolytic activity	1
		Decrease in ATP synthesis	91
Rabbit anti-bovine red cell AChE	unknown	Protection and restoration of enzymatic activity	55
Rabbit anti-Human red cell stroma	unknown	Loss of ATPase activity Decrease in K-transport	3
Ovine anti-M	unknown, antigen on HK sheep RBC*	?	77
			44
Ovine anti-L	unknown, antigen on LK sheep RBC*	Stimulation of active K-transport and S-ATPase in LK sheep red cells	16
			45
			46

*For explanation see text.

When cold-reacting anti-I antibody was used, the structural alterations appeared only in the cold and were reversed in the warm.[88] When anti-D, an antibody which did not agglutinate, was used, roughening of the surface and irregularity of the outline of the cell were seen, but the finger-like processes were not observed.

There are only a few observations available on the effects of antibody-antigen reactions at the red cell surface on biochemical processes within the membranes (Table 1). The red cell membrane contains a number of enzymes which might be affected by binding of antibodies to the red cell surface. Some of these enzymes have been demonstrated at the outer surface (e.g., acetyl-cholinesterase); some have been shown to function only at the inner surface of the red cell membrane (e.g., glycolytic enzymes). Others may be located within the membrane with binding sites for substrates and cofactors at the inside as well as at the outside of the membrane (Na$^+$-K$^+$ dependent ATPase), and their full function may depend entirely on the "sidedness" and structural integrity of the red cell membrane.

Acetylcholinesterase (AChE) is said to be located at the outer surface of the membrane[94] and has been shown to be an integral part of the membrane matrix.[42,58] The level of AChE is reported to be decreased in some instances of isoimmune hemolytic disease (e.g., ABO fetomaternal incompatibility)[7,37] but reactions in vitro of anti-A antibodies with normal red cells have failed to demonstrate depression of the AChE activity.[37] Even rabbit antisera produced to protein solubilized from human red cell membranes do not inhibit AChE activity.[43]

Michaeli et al.[55,56] found that an antibovine red cell AChE antibody obtained from rabbits stabilized the solubilized enzyme when the immune complex was exposed to 60°C. No change in substrate specificity could be observed. It was concluded that antigenic site and active site of the enzyme were remote from each other. When, however, the heat inactivated enzyme was mixed with the antibody, partial restoration of activity occurred in the soluble immune complex.[57] The renaturation did not restore the original conformation of the enzyme since a loss of substrate specificity for acetylthiocholine, but not for the less specific proprionylthiocholine, was observed. Although not performed on intact red cells or their membranes, these studies may present a model for the study of the interaction of antibody with an enzyme located at the outer surface of the membrane.

Enzymes of the anaerobic glycolytic system are located at the inner surface of the red cell membrane.[90] Since most of these enzymes do not remain in the white hemoglobin free ghost,[59] they are not assumed to be integral constituents of the membrane matrix. Studies on the action of antibodies on the glycolytic enzymes still attached to the membrane can, therefore, be done only on intact red cells. Anti-D (anti-Rh$_0$) has been shown to decrease the glycolytic rate in Rh$_0$-positive red cells and, furthermore, the blood of infants with erythroblastosis due to Rh incompatibility has been found to have a significant decrease of glycolytic activity.[1] If this effect of the anti-D antibody is not due to some interference or inhibition of substrate availability, then these antibodies probably exert their effect by changing the outer membrane structure and altering enzymatic activities at the inside of the membrane since antibodies probably do not penetrate to the inner membrane surface.

The viability of the erythrocyte depends in large part on the functional state of its active sodium (Na$^+$) and potassium (K$^+$) transport system, the so-called Na$^+$-K$^+$ pump. The location of this system within the membrane structure is not precisely defined; constituents of the outer as well as of the inner membrane probably participate in maintaining the activity of the Na$^+$-K$^+$ pump. Because of its interaction with high energy phosphates and cations at the inside and outside of the membrane, the Na$^+$-K$^+$ transport system is multifactorial and, in turn, depends on the structural integrity of the membrane. Thus, the Na$^+$-K$^+$ pump requires the presence of ATP as the energy source at the inner side of the membrane[14,32,112] to transport Na$^+$ from the inside to the outside and K$^+$ from the outside to the inside of the red cell against the electrochemical gradients of these ions. The active movement of cations maintains the steady state composition of the red cell and can be measured by net flux analysis, as well as by direct isotope labeling technics.[107] Active cation transport is inhibited by ouabain which competes with K$^+$ for a binding site at the outer surface of the membrane.[15,89] Less than 10 per cent of the glycolysis-produced free energy (in the form of ATP) has been found to be necessary to drive the Na$^+$-K$^+$ pump in human red cells.[102] Further, a Na$^+$-K$^+$ stimulated and ouabain sensitive ATPase (S-ATPase), first described by Skou[99,100] in crab nerves has also been discovered in red cells.[74] A correlation between both processes, the active Na$^+$-K$^+$ transport and the S-ATPase activity, has been suggested. The full function of

S-ATPase depends upon the availability of ATP at the inner side of the membrane, as well as on the "sidedness" of the membrane with respect to cation concentrations. The action of antibodies on the Na^+-K^+ transport system and the associated ATPase must, therefore, be seen in the light of their complicated function and differentiated structural organization in the membrane matrix. Thus, the biosynthesis of ATP and its binding and hydrolysis by ATPase, as well as binding and transport of cations across the membrane, may be affected separately or as a complex by the action of antibodies.

A few observations on the action of some blood group antibodies on ATP biosynthesis and ATPase activity in human red cells have been reported. Schrier et al[91] found that sensitization of red cells containing the D (Rh_0) antigen with hyperimmune anti-D serum or its IgG fraction induced a more than 30 per cent decrease in ATP synthesis, as determined by the membrane-mediated incorporation of radioactive inorganic phosphate into ATP. No effect was found with cells lacking the D (Rh_0) antigen, nor with anti-A serum on type O D-positive and D-negative red cell membranes. The observed effect could be related neither to impaired substrate availability nor to increased ATP destruction. The site of inhibition has not yet been defined.

Palek et al.[70] studied the effect of isoimmune anti-A serum on the level of organic phosphates in human type A red cells. They found a decrease in total ATP content associated with an increase in the ADP and inorganic phosphate (P_i) content only in A_1-red cells treated with anti-A, but not in A_1 cells treated with autologous serum. This observation was paralleled by the finding of an increase in total ATPase activity in membranes obtained from A_1 red cells treated with anti-A. No data, however, were given to indicate whether or not anti-A affects the Na^+-K^+ dependent ATPase in human red cell membranes, and if so, whether or not a similar effect could also be produced with the same antiserum using O and B type red cells. Because of the role of the Na^+-K^+ dependent S-ATPase for maintaining the steady state composition of the red cell, examination of its activity will be crucial in evaluating antibody action on the ATPase system.

The source of the antiserum used to study the action of antibodies on membrane function (e.g., the ATPase) is of importance. Recently, Averdunk et al.[3] reported that antihuman red cell membrane antibodies prepared in rabbits inhibited Na^+-K^+ transport and S-ATPase activity in human red cells and their membranes, respectively. Using net flux measurements, it was demonstrated that intact red cells showed a net loss of K^+ and a net gain of Na^+ as a function of the amount of antiserum added to the cells. The effect appeared to be specific since no change in glucose consumption was observed. In addition, incubation of white membranes with the heterologous antiserum inhibited both the ouabain sensitive S-ATPase and the ouabain insensitive Mg-ATPase (I-ATPase) with a concomitant drop of the S-ATPase/I-ATPase ratio from 1.3 to less than 0.2. Heterologous *non-*immune serum did not produce either of these effects. It should be noted that rabbit antisera against human red cell membranes may contain antibodies for the carbohydrate moieties (e.g., Forssman-like antigens) and also contain a considerable number of antibodies reacting with the membrane proteins extracted from the red cell membrane by various organic solvents.[75] The specificity of antibodies causing these effects remains to be elucidated.

Recently, new aspects of the action of antibodies on red cell membrane function have come from studies of the effect of ovine iso-immune sera on cation transport in high potassium (HK) and low potassium (LK) sheep red cells. The inheritance of the HK and LK character follows simple Mendelian rules whereby the LK gene appears to be dominant.[17] The difference in the steady state composition of the HK and LK cells can be attributed to differences in active and passive Na^+-K^+ transport.[108] In particular, the active transport of these cations as well as the Na^+-K^+ stimulated and ouabain-sensitive ATPase activity are about four times higher in HK than in LK sheep red cells,[108,109] suggesting a close association of the two processes.[110] Using ovine isoimmune antisera, Rasmusen and Hall[77] discovered that the inheritance of an antigen designated M is associated with the HK character of sheep red cells. Subsequently, Rasmusen[76] and Ellory and Tucker[16] have described another antigen called L which is found only on LK sheep red cells. The presence of both antigens in the heterozygous state indicates a closed genetic system. In the presence of complement, anti-M lyses only HK cells and the red cells of M-positive LK sheep, while anti-L lyses only LK cells. The correlation between cation type and antigenic type in HK and LK sheep red cells is shown in Table 2.

Lauf and Tosteson[44] showed that the amount and affinity of anti-M antibody bound per HK cell or membrane was not changed in the presence of ATP and ions required for the function of the Na^+-K^+ dependent ATPase activity. Addition of ouabain inhibited the S-ATPase and decreased the total ATPase activity to less than one-fourth of its original value. Since the binding of the anti-M antibody was unchanged in the presence of ouabain, it was concluded that the M-antigenic site and the active site of the enzyme may be structurally separate from each other.[44] This conclusion was supported by the observation that anti-M serum had no effect on the S-ATPase activity[6,16,45] in HK sheep red cell membranes and did not alter significantly the K^+-pump influx in intact HK red cells.[45]

A clearer functional relationship was found when the interaction of LK cells with the L-antiserum prepared by immunizing HK sheep with LK blood was investigated. In confirmation of the observations of Ellory and Tucker,[16] Lauf et al.[46,47] found that incubation of homozygous LK sheep red cells with anti-L antiserum increased K^+ influx in LK cells about four to five fold but did not affect the Na^+-K^+ transport in HK sheep red cells. This effect could also be demonstrated by using diluted and specifically absorbed anti-L antiserum, thus precluding the possibility of interference by antibodies specific for antigens other than the L-antigen. Potassium pump stimulation was also seen when the IgG_1 protein isolated from the L-

Table 2.—Relation of Cation Type and Antigens on Sheep Red Cells*

Cation Type	Antigens on Red Cells†	
	M-antigen	L-antigen
High potassium (HK) (homozygous)	+	—
Low potassium (LK) (heterozygous)	+	+
Low potassium (LK) (homozygous)	—	+

*From Refs. 76, 77, 16, 44, and 46.
†As detected by hemolysis in presence of complement.

antiserum by ion exchange chromatography was tested. The number of Na$^+$-K$^+$ pump sites per cell may be estimated by measuring the uptake of tritiated ouabain molecules per cell,[33] assuming a stoichiometric ratio between the number of pump sites and the number of ouabain binding sites. Using this method, it was found that anti-L treated LK cells had a two-fold increase in the number of pump sites per cell as compared to the nonimmune serum treated control.[47] This twofold increase in the number of pump sites did not fully account for the four to six fold stimulation of the maximum pumping rate observed. Kinetic studies indicated that the antigen-antibody reaction resulted in the appearance of new pump sites closer in their character to those of HK cells than to those of LK cells. However, the affinity of the anti-L treated LK cells to external K$^+$ was unchanged, a finding which was somewhat incompatible with this concept, and further experiments will be necessary for clarification of this point.

In addition to these flux studies, it is known that the anti-L antiserum stimulates several fold the S-ATPase activity in LK red cell membranes.[16] Membranes from LK red cells have only a very small fraction of S-ATPase activity as compared to HK membranes, and the S-ATPase/I-ATPase ratio appears to be around 0.2 under optimal conditions. Incubation of LK membranes with the anti-L antiserum stimulates the S-ATPase without a change in total ATPase activity, thus increasing the S-ATPase/I-ATPase ratio. The stimulation of the S-ATPase is, however, not of the consistency that is found for the stimulation of the K-pump influx.[48] It is possible that differences in the effect brought about by anti-L may be related to several L-alleles of the L-antigen or to the state of the membranes following release of hemoglobin by osmotic lysis.

EFFECT OF COMPLEMENT ON RED CELL MEMBRANE

Serum complement is the name given to a system of nine proteins which react sequentially and which account for many of the effects of immune reactions, including the inflammatory response, leukotaxis, and cytolysis.[64] The complement sequence is most efficiently initiated by the activation of antibody at the cell surface. The reaction of antibody with antigen probably alters the molecule, facilitating reaction with the first component of complement (C1). All studies to date have suggested that two such interactions with antibody are required for the fixation of C1 and that these can either be supplied by two IgG molecules or by two subunits of a single IgM molecule.[5,83]

C1 is a complicated molecule consisting of three protein units held together by calcium.[49] The first part, C1q, is the portion which interacts with antibody.[65] This interaction activates a second portion, C1r, which in turn activates the third portion. C1s, from a proesterase to an esterase.[68] The C1 molecule does not appear to be closely attached to the red cell membrane since, when the antibody is eluted, C1 is likewise removed.[84] No effects of C1 on the structural, physical, or biochemical characteristics of the membrane are known.

One function of C1 is to activate C4 by an esterolytic reaction in which the C4 molecule is "opened up" and is firmly attached to membrane receptors.[64] The exact nature of the membrane receptors is not certain, but C4 may not be easily removed from the membrane. The effect of the presence of C4 on the membrane has not been rigorously studied.

When the next component of complement, C2, is acted upon by C1, the molecule is activated and firmly attached to C4.[24,98,106] The combination of C2 and C4 functions as an enzyme which splits and activates the next component, C3, a portion of which is firmly attached to the red cell membrane.[67,96] It is thought that a single C4-C2 site may bring about the attachment of anywhere from 8 or 10[96] to several hundred C3 molecules.[67] Although C3 molecules may accumulate on the red cell surface,[30] the effect of C3 on the red cell membrane has not been completely studied. To date, no evidence of distinct morphologic or biochemical changes has been noted.

C3 apparently acts as a peptidase upon the next component, C5, splitting the molecule.[10,11,96] A small, 15,000 molecular weight piece is released into the fluid phase; this material has a potent leukotactic activity.[95] The larger part of the molecule is firmly attached to the red cell where it participates in further complement reactions. The reactions involved in the fixation of the next four components, C6, C7, C8, and C9, are not well studied to date since the components are only now being purified.

The last two components, C8 and C9, are involved in the final cytolytic steps of the complement sequence. Cytolysis results after completion of at least two steps after the addition of C9 to the cell. One step is inhibited by reduced temperature, another by high concentrations of ethylenediaminetetraacetate (EDTA).[22,23] The final result of these reactions is the formation of the membrane defect, 80–100 Ångstroms in diameter (Fig. 1).[35] When viewed in electron microscopy with negative stain, the defect appears to be surrounded by a ring, possibly of membrane material. The size and morphology of the primary defect appears to be independent of the membrane surface affected since the same membrane defects appear in bacterial membranes, red cell membranes, membranes of nucleated cells and artificial lipoprotein membranes acted upon by complement. The size of the primary membrane defect likewise does not appear to be related to the source and type of antibody but may differ with different complement species. The membrane defect formed by human complement appears to be somewhat

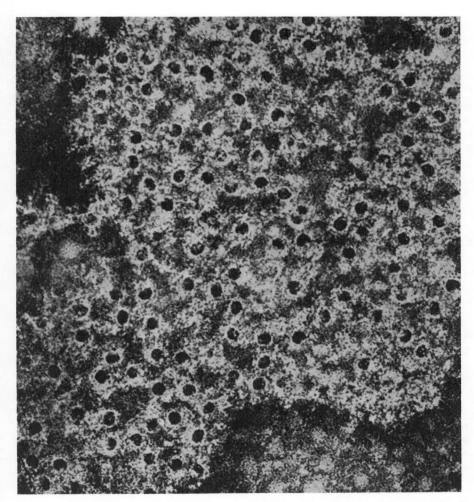

Fig. 1.—Membrane defects formed by excess human complement on human red cells sensitized with anti-I, as shown with phosphotungstate negative stain ($\times$ 374,000, taken by Dr. R. R. Dourmashkin).

larger than that formed by guinea pig and rabbit complement.[78]

The number of defects produced in the red cell membrane varies with different conditions of lysis. According to the "one-hit" theory of complement lysis, which assumes that a single completed complement sequence is sufficient to lyse the cell, a mean of one complement sequence is completed when 63 per cent of the cells are hemolyzed.[54] If each sequence results in the formation of one hole, then a mean of one hole per cell would be seen at this level of lysis. In sheep cells lysed by IgM anti-Forssman antibody and guinea pig complement, Borsos et al.[4] have shown that the number of membrane defects predicted by this calculation appears when limited amounts of lysis are allowed to take place. When human complement is used to lyse

the cells, a vastly greater number of membrane defects appears per cell
so that, when 10 per cent of the membranes are lysed by human complement,
a mean of 10,000 defects are counted.[78] These defects appear only in the
lysed cells and are not present on the unlysed cells. Because of the peculiar
resistance of human red cells to lysis by complement, very high concentrations of human serum are required to obtain lysis of this percentage of
normal cells. When equally high concentrations of human serum as a source
of C3–C9 are used on sheep cells, a similarly large number of membrane
defects appears.[35] The excessive number of holes is not seen with guinea-
pig serum, except when the antibody used to sensitize the sheep cell is an
IgM and lysis is limited by antibody concentration, complement being present
in excess. If lysis is limited by complement concentration, the "multiple
hole" effect is not seen with either human or guinea-pig complement.

Several questions remain concerning the multiple hole effect. It has been
suggested that because of the numerical discrepancy, these defects do not
represent through-and-through holes.[31] However, the fact that they do not
appear on the unlysed cells would suggest that they are not holes that have
been repaired. The completeness of the defect remains to be seen using ion
flux studies. It is possible that once a hole is formed, the membrane is so
changed that other holes are formed more readily. Alternatively, a hypothetical lytic agent generated by C9 might be transferred to other sites on the
membrane. The effect remains a mystery.

How the membrane defect is formed is unknown. It has been suggested
that C8 and/or C9 is a lipase and that this in some way alters the membrane. Fischer and Haupt[21] have suggested that the membrane defect is
brought about by the deacylation of lecithin, forming lysolecithin locally
within the membrane and thus leading to its disruption. However, there is
considerable evidence to the contrary. Smith and Becker[101] have found
that lysolecithin cannot be detected in membranes lysed with complement.
Further, the morphological alteration in the membrane effected by lysolecithin is very different from that produced by complement.[13]

The complement-dependent defects appear similar to those formed by
saponin with the exception that the defects formed by saponin appear to
be more regular in array and tend to be somewhat more hexagonal than
the complement defects.[13] The saponin molecule has a lipophilic and a
lipophobic portion. It has been suggested that the saponin membrane defects
appear because the lipophilic portion of the molecule complexes with membrane cholesterol and the water-soluble hydrohilic portion is then directed
toward the center of the defect.[52] To date, no certain evidence of the nature
of the molecule responsible for the complement-dependent defect has been
obtained.

The major effect of the disruption of the membrane surface is the pertur-
bation of the ionic equilibrium of the cell. The Na^+-K^+ pump which maintains
the differential cationic concentrations is essentially bypassed with the result
that potassium very rapidly leaves the cell.[28,31,93] The rate at which potassium leaves the cell is proportional to the number of membrane defects

formed on the cell.[31] Since the size of the primary lesion is apparently insufficiently large to permit the egress of hemoglobin, cell lysis (as determined by the presence of hemoglobin outside the cell membrane) does not occur immediately. Once the membrane has been breached, the usual steady, stable state no longer obtains and the concentration of ions on both sides of the membrane presumably becomes equal. The hemoglobin contained within the cell continues to exert an osmotic force, drawing water into the cell which results in osmotic swelling. Disruption of the membrane or leak of the hemoglobin through distorted membrane defects then occurs and cell lysis is signalled by the release of hemoglobin. The swelling and disruption of the cell can be prevented by the presence in the outside medium of an osmotic macromolecule unable to penetrate the defect and equal in concentration to the concentration of hemoglobin within the cell.[22,28,93] However, if many defects are formed, hemoglobin is able to leak directly and the presence of albumin or other macromolecules in the outside medium does not prevent hemolysis.[93]

This lytic mechanism (colloid-osmotic lysis) appears to be unique for red cells. When other cells are affected by complement, there is a loss of cation and a loss of the small molecular weight materials, such as RNA.[28] The concentration of osmotically active material retained within the cell may not be sufficient to bring about lysis by colloid-osmotic swelling. Once the defect has occurred in the cell membrane, the cell is no longer able to exclude dyes; this is the basis for the trypan blue exclusion technic for the detection of complement lysis of nucleated cells.

The efficiency with which the complement sequence is able to effect lysis of the red cell by formation of the membrane defect appears to depend upon both the concentration of complement components and upon the nature of the membrane acted upon. If insufficient amounts of any one of the complement components are present, the reaction cannot proceed and lysis cannot occur. However, the completion of a single complement sequence in the formation of a single membrane defect is probably sufficient for the destruction of the red cell. This thesis has been established for the sheep red cell-rabbit antibody-guinea pig complement system by the work of Mayer and his group.[54] It has not been rigorously proved for human cells, but it is difficult to imagine how a cell can survive the ionic imbalance consequent to the presence of even a single membrane defect.

The efficiency with which the complement sequence is completed varies markedly from cell to cell. For the sheep cell, 70 to 80 per cent of all complement sequences which are initiated are able to continue to completion, if all components are present in adequate concentrations.[9] The same is not true for human cells and it appears that on the normal human red cell, less than one initiated sequence in 200 is able to go through to completion and to result in lysis.[79] The reason for this discrepancy is not apparent. A variable proportion of the red cells of patients with paroxysmal nocturnal hemoglobinuria (PNH), on the other hand, are much more sensitive to the lytic action of complement.[80,81] This difference does not appear to be due to

an increased uptake of at least the first four components[82] but does appear
to be due to an increase in the efficiency with which each sequence, once
begun, proceeds through to completion. This increase in efficiency is presumably due to a defect in the membrane, probably in the membrane proteins,
and can be mimicked by pretreatment of the red cells with certain sulfhydryl
reagents.[12,36,97] The nature of the membrane defect and the way in which
it is able to make complement reaction more efficient is not known. It does
not appear that the efficiency of complement sequences, even on PNH
cells, anywhere near approaches that on sheep cells.

EFFECTS OF IMMUNE REACTIONS ON RED CELLS ON LYSIS IN VIVO

In the above sections, we have outlined the demonstrable effects of immune
reactions carried out in vitro. Precisely how, if at all, these reactions relate
to the survival of the red cell in vivo is not, in all instances, clearly understood.

Red cells may be destroyed during their circulation by the presence of
antibody alone since some antibodies (e.g., antibodies in the Rh system)
do not fix complement.[30,84] In these instances, the rate at which the red
cell is destroyed is in large part dependent upon the concentration of antibody on the red cell surface.[60] This relationship has been shown by determining the survival of red cells artifically sensitized with known amounts of
anti-D antibody. The rate of destruction upon reinjection is related to the
amount of antibody added to the cells. If the amount of antibody is small,
these cells are largely removed in the spleen. When larger amounts of antibody are present on the cell surface, the cells are removed by both the
liver and the spleen.

The same relationship appears to hold between the amount of antibody
present on the cells and their rate of destruction in patients with autoimmune hemolytic anemia.[85] Up to several hundred antibody molecules may
be present on the red cell surface without a significant decrease in red cell
survival time. However, as the amount of antibody present on the red cell
is increased, the rate of hemolysis is increased in proportion to the increase
in antibody concentration. In patients who have undergone splenectomy,
the amount of antibody on the red cell may be markedly increased and the
survival of the cells may be nearly normal. However, when an even greater
amount of antibody is present on the cell, the cell is destroyed despite the
absence of the spleen.

The mechanism by which cells sensitized by IgG antibody alone are
destroyed apparently does not have to do with an effect on the red cell
membrane itself. Although hemagglutination may occur, it is unknown
whether or not this has a deleterious effect on the survival of the cell. The
main mechanism of destruction probably is the sequestration of the cell
due to the interaction of the antibody coating the red cell with macrophages
of the reticuloendothelial system, especially in the spleen. Based on the earlier
work of Policard and Bessis,[72] as well as the recent studies of Archer[2] and
LoBuglio et al.[51] on macrophage interaction with antibody coated red cells,

it has become clear that phagocytosis of just a portion of the red cell occurs. This process causes the unphagocytosed portion of the cell to become increasingly spherical and this change probably accounts for the spherocytosis seen in immune hemolytic states.

When the complement sequence is initiated by the antibody on the cell, lysis in vivo may result from the formation of membrane defects upon completion of the sequence. This process leads to hemoglobinemia and, usually, hemoglobinuria and hemosiderinuria. In patients with paroxysmal cold hemoglobinuria, most of the lysis is probably by this mechanism. The potent lytic ability of the Donath-Landsteiner antibody is, as yet, unexplained. Other complement fixing antibodies may cause lysis by this mechanism as well, especially some iso-antibodies encountered in transfusion reactions. On the other hand, it is not apparent whether or not *all* lysis due to complement is by this mechanism. As noted above, not all complement sequences once begun on human red cells go through to completion, and the cells of patients with complement fixing antibodies very frequently contain large amounts of C4 and C3.[18,30,71] This finding has led to the supposition that there may be another mechanism for the destruction of the cells by means of the interaction between the complement component coating and some part of the reticuloendothelial system.

The presence of C3 on the cell surface renders the cell susceptible to erythrophagocytosis by both polymorphonuclear leukocytes and macrophages in vitro.[69] The relationship of this phenomenon to cell survival in vivo has been questioned since the C3-covered cells of patients with cold-agglutinin disease are not subject to rapid destruction during circulation, whether the C3 has been added in vivo[19] or in vitro.[50] The question of the role of the presence of complement components on the red cell membrane in shortening the red cell life span, remains open.

In the foregoing review, we have tried to outline what is known and what is not known about the effects of immune reactions on the red cell membrane and the significance of these effects for the survival of the cell. Insufficient knowledge is available concerning the role of the antigens on the red cell surface so that predicted consequences of antigen-antibody interaction are difficult to ascertain. Antigen-antibody reactions at the red cell surface can, in some instances, increase or depress certain membrane functions, but considerably more work is necessary to determine the range of these reactions and their effect on cell survival. The mechanism by which complement brings about the destruction of the red cell is understood somewhat better, but considerable gaps of knowledge remain concerning the relationship of this change to biochemical events on the cell surface on the one hand and to the destruction of the cell in vivo on the other.

REFERENCES

1. Abraham, A., and Diamond, L. K.: Erythrocyte glycolysis in erythroblastotic newborns. Amer. J. Dis. Child 99:202, 1960.

2. Archer, G. T.: Phagocytosis by human monocytes of red cells coated with Rh antibodies. Vox Sang. 10:590, 1965.

3. Averdunk, R., Gunther, T., Dorn, F., and Zimmermann, U.: Uber die Wirkung

von Antikorpern auf die ATPase Aktivitat und Menschen-Erythrozyten. Z. Naturforsch. und den aktiven Na-k-Transport von E. coli 24b:693, 1969.

4. Borsos, T., Dourmashkin, R. R., and Humphrey, J. H.: Lesions in erythrocyte membranes caused by immune haemolysis. Nature 202:251, 1964.

5. —, and Rapp, H. J.: Hemolysin titration based on the fixation of the activated first component of complement: evidence that one molecule of hemolysin suffices to sensitize an erythrocyte. J. Immun. 95:559, 1965.

6. Brewer, G. J., Eaton, J. W., Beck, C. C., Feitler, L., and Shreffler, C. C.: Sodium-potassium stimulated ATPase activity of mammalian hemolysates: clinical observations and dominance of ATPase deficiency in the potassium polymorphism of sheep. J. Lab. Clin. Med. 71:744, 1968.

7. Burman, D.: Red cell cholinesterase in infancy and childhood. Arch. Dis. Child. 36:362, 1961.

8. Cohen, S., and Milstein, C.: Structure and biological properties of immunoglobulins. Advances Immun. 8:1, 1968.

9. Colten, H. R., Borsos, T., and Rapp, H. J.: Efficiency of the first component of complement (C'1) in the hemolytic reaction. Science 158:1590, 1967.

10. Cooper, N. B., and Becker, E. L.: Complement-associated peptidase activity of guinea pig serum. I. Role of complement components. J. Immun. 98:119, 1967.

11. —: Complement-associated peptidase activity of guinea pig serum. II. Role of a low molecular weight enhancing factor. J. Immun. 98:132, 1967.

12. De Sandre, G., Vettore, L., Corrocher, R., Cortesi, S., and Perona, G.: Ham-positive red cells induced in vitro by N-acetylcysteine or D-penicillamine. Brit. J. Haemat. 15:437, 1968.

13. Dourmashkin, R. R., and Rosse, W. F.: Morphologic changes in the membranes of red blood cells undergoing hemolysis. Amer. J. Med. 41:699, 1966.

14. Dunham, E. T.: Linkage of active transport to ATP utilization. Physiologist 1:23, 1957.

15. —, and Glynn, I. M.: Adenosinetriphosphatase activity and the active movements of alkali metal ions. J. Physiol. (London) 156:274, 1961.

16. Ellory, I. C., and Tucker, E. M.: Stimulation of the potassium transport system in low potassium type sheep red cells by a specific antigen-antibody reaction. Nature 222:477, 1969.

17. Evans, J. V.: Electrolyte concentrations in red blood cells of British breeds of sheep. Nature 174:931, 1954.

18. Evans, R. S., Turner, E. T., and Bingham, M.: Chronic hemolytic anemia due to cold agglutinins: the mechanism of resistance of red cells to C' hemolysis by cold agglutinins. J. Clin. Invest. 46:1461, 1967.

19. —, —, —, and Woods, R.: Chronic hemolytic anemia due to cold agglutinins. II. The role of C' in red cell destruction. J. Clin. Invest. 47:691, 1968.

20. Eylar, E. H., Madoff, M. A., Brody, O. V., and Oncley, J. L.: The contribution of sialic acid to the surface charge of the erythrocyte. J. Biol. Chem. 237:1992, 1962.

21. Fischer, H., and Haupt, I.: Das cytolisierende Prinzip von Serumkomplement. Z. Naturforsch. 16b:321, 1961.

22. Frank, M. M., Rapp, H. J., and Borsos, T.: Studies on the terminal steps of immune hemolysis. II. Resolution of the E* transformation reaction into multiple steps. J. Immun. 94:295, 1965.

23. —, Rapp, H. J., and Borsos, T.: Studies on the terminal steps of immune hemolysis. I. Inhibition by trisodium ethylenediaminetetraacetate (EDTA). J. Immun. 93:409, 1964.

24. Gigli, I., and Austen, F. K.: Fluid phase destruction of C2^{hu} by C1^{hu}. I. Its enhancement and inhibition by homologous and heterologous C4. J. Exp. Med. 129:679, 1969.

25. Glaser, R. M., and Mel, H. C.: The electrophoretic behavior of osmium tetroxide-fixed and potassium permanganate-fixed rat erythrocytes. Biochim. Biophys. Acta. 79:606, 1963.

26. —, and —: Microelectrophoretic and enzymic studies concerning the carbohydrate at the surface of rat erythrocytes. Arch. Biochem. 113:77, 1966.

27. Glynn, I. M.: The action of cardiac glycosides on sodium and potassium movements in human red cells. J. Physiol. (London) 136:148, 1957.

28. Green, H., Barrow, P., and Goldberg, B.: Effect of antibody and complement on permeability control in ascites tumor cells and erythrocytes. J. Exp. Med. 110:699, 1959.

29. Hakamori, S., and Jeanloz, R. W.:

Isolation and characterization of glycolipids from erythrocytes of human blood A (plus) and B (plus). J. Biol. Chem. 236:2827, 1961.

30. Harboe, M., Müller-Eberhard, H. J., Fudenberg, H., Polley, M. J., and Mollison, P. L.: Identification of the components of complement participating in the antiglobulin reaction. Immunology 6:412, 1963.

31. Hingson, D. J., Massengill, R. K., and Mayer, M. M.: The kinetics of release of [86]rubidium and hemoglobin from erythrocytes damaged by antibody and complement. Immunochemistry 6:295, 1969.

32. Hoffman, J. F.: The active transport of sodium by ghosts of human red blood cells. J. Gen. Physiol. 45:837, 1962.

33. —, and Ingram, C. J.: Cation transport and the binding of T-ouabain to intact human red blood cells. In Deutsch, E., Gerlach, E., and Moser, K., (Eds.): Metabolism and Membrane Permeability of Erythrocytes and Thrombocytes. Stuttgart, Georg Thieme, 1968, p. 420.

34. Hoyer, L. W., and Trabold, N. C.: The significance of erythrocyte antigen site density. I. Hemagglutination. J. Clin. Invest. 49:87, 1970.

35. Humphrey, J. H., and Dourmashkin, R. R.: The lesions in cell membranes caused by complement. In Dixon, F. J., Jr., and Kunkel, H. G. (Eds.): Advances in Immunology, Vol. II. New York, Academic Press, 1969, p. 75.

36. Kann, H. E., Jr., Mengel, C. E., Meriwether, W. D., and Ebbert, L.: Production of in vitro lytic characteristics of paroxysmal nocturnal hemoglobinuria erythrocytes in normal erythrocytes. Blood 32:99, 1969.

37. Kaplan, E., Herz, G., and Hsu, K. S.: Erythrocyte acetylcholinesterase activity in ABO hemolytic disease of the newborn. Pediatrics 33:205, 1964.

38. Kaplan, M. E., and Kabat, E. A.: Studies on human antibodies. IV. Purification and properties of anti-A and anti-B obtained by absorption and elution from insoluble blood group substance. J. Exp. Med. 123:1061, 1966.

39. Klenk, E., and Uhlenbruck, G.: Über neuraminsäurehaltige Mucoide aus Menschenerythrozytenstroma. Ein Beitrag zur Chemie der Agglutinogene. Hoppe Seyler Z. Physiol. Chem. 319:151, 1960.

40. Koscielak, J.: Blood group A specific glycolipids from human erythrocytes. Biochim. Biophys. Acta 78:313, 1963.

41. Landsteiner, K.: Zur Kenntnis der antifermentativen, lytischen und agglutinierenden Wirkungen des Blutserums und der Lymphe. Zbl. Bakt. 27:357, 1900.

42. Lauf, P. K., and Poulik, M. D.: Solubilization and structural integrity of the human red cell membrane. Brit. J. Haemat. 15:191, 1968.

43. —: Unpublished results.

44. —, and Tosteson, D. C.: The M-antigen in HK and LK sheep red cell membranes. J. Membrane Biol. 1:177, 1969.

45. —, Rasmusen, B. A., and Tosteson, D. C.: Action of isoimmune anti-M and anti-L on active potassium transport in HK and LK sheep red cells. In Aminoff, D. A. (Ed.): Blood and Tissue Antigens. New York, Academic Press, 1970, p. 341.

46. —, —, Hoffman, P. G., Dunham, P. B. Cook, P. H., Parmelee, M. L., and Tosteson, D. C.: Stimulation of active K+-transport in LK sheep red cells by anti-L serum. Third Int. Congress Int. Union Pure and Applied Biophysics. Cambridge, Mass., 1969, p. 69.

47. —, —, —, —, —, —, and —: Stimulation of active potassium transport in LK sheep red cells by blood group-L antiserum. J. Membrane Biol. 3:1, 1970.

48. —, Parmelee, M. L., and Snyder, J. J.: Stimulation of S-ATPase in LK sheep red cell membranes by isoimmune anti-L serum. 14th Annual Meeting Biophysical Society, Baltimore, 1970. Abstract. 10:7a.

49. Lepow, I. H., Naff, G. B., Todd, E. W., Pensky, J., and Hinz, C. F.: Chromatographic resolution of the first component of human complement into three activities. J. Exp. Med. 117:983, 1963.

50. Lewis, S. M., Dacie, J. V., and Szur, L.: Mechanism of haemolysis in the cold-haemagglutinin syndrome. Brit. J. Haemat. 6:154, 1960.

51. LoBuglio, A. F., Cotran, R. S., and Jandl, J. H.: Red cells coated with immunoglobulin G: binding and sphering by mononuclear cells in man. Science 158:1582, 1967.

52. Lucey, J. A., and Galuert, A. M.: Structure and assembly of macromolecular lipid complexes composed of globular micelles. J. Molec. Biol. 8:727, 1964.

53. Marcus, D. M., and Cass, L. E.: Glycosphingolipids with Lewis blood group

activity: uptake by human erythrocytes. Science 164:553, 1969.

54. Mayer, M. M.: Complement and complement fixation. *In* Kabat, E. A. and Mayer, M. M., (Eds.): Experimental Immunochemistry, 2nd Ed. Springfield, Ill., Charles C Thomas, 1961, pp. 133-235.

55. Michaeli, D., Pinto, J. D., and Benjamini, E.: Restoration of enzyme activity of heat-denatured acetylcholinesterase by antibodies to the native enzyme. Nature 213:77, 1967.

56. —, —, —, and de Buren, F. P.: Immunoenzymology of acetylcholinesterase. I. Substrate specificity and heat stability of acetylcholinesterase and of acetylcholinesterase antibody complex. Immunochemistry 6:101, 1969.

57. —, —, and —: Immunoenzymology of acetylcholinesterase. II. Effect of antibody on the heat-denatured enzyme. Immunochemistry 6:371, 1969.

58. —, and Hanahan, D. J.: Solubilization of certain proteins from the human erythrocyte stroma. Biochemistry 5:51, 1966.

59. —, Mitchell, W. B., and Hanahan, D. J.: Enzyme and hemoglobin retention in human erythrocyte stroma. Biochim. Biophys. Acta 104:348, 1965.

60. Mollison, P. L., Crome, P., Hughes-Jones, N. C., and Rochna, E.: Rate of removal from the circulation of red cells sensitized with different amounts of antibody. Brit. J. Haemat. 11:461, 1965.

61. Moreno, C., and Kabat, E. A.: Studies on human antibodies. VIII. Properties and association constants of human antibodies to blood group A substance purified with insoluble specific adsorbents and fractionally eluted with mono- and oligosaccharides. J. Exp. Med. 129:871, 1969.

62. Morgan, W. T. J., and Watkins, W. M.: The inactivation of the blood group receptors on the human erythrocyte surface by the periodate ion. Brit. J. Exp. Path. 32:34, 1951.

63. Morgan, W. R. J.: Chemische Grundlagen der menlischen Blutgruppenspezifitat. Naturwissenschaften 46:181, 1959.

64. Müller-Eberhard, H. J.: Chemistry and Reaction Mechanisms of Complement. New York, Academic Press, 1968. Vol. 8, pp. 1-80.

65. Müller-Eberhard, H. J., and Kunkel, H. G.: Isolation of a thermolabile serum protein which precipitates γ-globulin aggre-

gates and participates in immune hemolysis. Proc. Soc. Exp. Biol. Med. 106:291, 1961.

66. —, and Lepow, I. H.: C'1 esterase effect on activity and physicochemical properties of the fourth component of complement. J. Exp. Med. 121:819, 1965.

67. —, Dalmasso, A. P., and Calcott, M. A.: The reaction mechanism of $\beta 1_c$-globulin (C'3) in immune hemolysis. J. Exp. Med. 123:33, 1966.

68. Naff, G. B., and Ratnoff, O. D.: The enzymic nature of C'1r conversion of C'1s to C'1 esterase and digestion of amino acid esters by C'1r. J. Exp. Med. 128:571, 1968.

69. Nelson, R. Á., Jr.: Immune adherence. *In:* Grabar, P., and Miescher, P. (Eds.): Mechanism of Cell and Tissue Damage Produced by Immune Reactions. Basel, Schwabe & Co., 1962, p. 245.

70. Palek, J., Mircevova, L., Brabec, V., Friedmann, B., and Majsky, A.: The effect of anti-A antibody on red cell organic phosphates and adenosine triphosphatase activity in vitro. Scand. J. Haemat. 5:191, 1968.

71. Peetoom, T., and Pondman, K. W.: The significance of antigen-antibody complement reactions. III. The identification of C'4 in the immunoelectrophoretic pattern obtained with anti-human complement serum. Vox Sang. 8:605, 1963.

72. Policard, A., and Bessis, M.: Fractionnement d'hématies par les leucocytes au course de la phagocytose. C. R. Soc. Biol. (Paris) 147:982, 1953.

73. Pollack, W., Hager, H. J., Reckel, R., Toren, D. A., and Singher, A.: A study of the forces involved in the second stage of hemagglutination. Transfusion 5:158, 1965.

74. Post, R. L., Merritt, C. R., Kinsolving, C. R., and Albright, C. D.: Membrane adenosine triphosphatase as a participant of active transport of sodium and potassium in the human erythrocyte. J. Biol. Chem. 235:1796, 1960.

75. Poulik, M. D., and Lauf, P. K.: Immunological properties of "soluble" proteins of human red cell ghosts. Fed. Proc. 28:315, 1969.

76. Rasmusen, B. A.: A blood group antibody which reacts exclusively with LK sheep red blood cells. Genetics 61:49, 1969.

77. —, and Hall, J. G.: Association between potassium concentration and serological type of sheep red blood cells. Science 151:1551, 1966.

78. Rosse, W. F., Dourmashkin, R., and

Humphrey, J. R.: The lysis of normal and paroxysmal nocturnal hemoglobinuria (PNH) red blood cells by antibody and complement. III. The membrane defects produced by complement lysis. J. Exp. Med. 123:969, 1966.

79. —, and Jenkins, J. H.: Unpublished observations.

80. —, and Dacie, J. V.: The lysis of normal and paroxysmal nocturnal hemoglobinuria (PNH) red blood cells by antibody and complement. I. The sensitivity of PNH red cells to lysis by complement and specific antibody. J. Clin. Invest. 45:736, 1966.

81. —, and —: The lysis of normal and paroxysmal nocturnal hemoglobinuria (PNH) red blood cells by antibody and complement. II. The relative role of complement and antibody in the sensitivity of PNH cells to lysis. J. Clin. Invest. 45:749, 1966.

82. —, and Logue, G. L.: The efficiency of complement reactions on normal and paroxysmal nocturnal hemoglobinuria red cells. J. Clin. Invest. 1970, In press.

83. —, Borsos, T., and Rapp, H. J.: Structural characteristics of hemolytic antibodies as determined by the effects of ionizing radiation. J. Immun. 98:1190, 1967.

84. —: Fixation of the first component of complement (C'1a) by human antibodies. J. Clin. Invest. 47:2430, 1968.

85. —: Quantitative antibody studies in autoimmune hemolytic anemia due to warm-reactive antibodies. J. Clin. Invest. 224:70a, 1969.

86. Ruhenstroth-Bauer, G.: The electric charge of blood cells. In Sawyer, P. N. (Ed.): Biophysical Mechanism in Vascular Homeostasis and Intravascular Thrombosis. New York, Appleton Century Crofts, 1965, p. 42.

87. Salsbury, A. J., and Clarke, J. A.: Surface changes in red blood cells undergoing agglutination. Rev. Franc. Etud. Clin. Biol. 12:981-986, 1967.

88. —, —, and Shaul, W. S.: Red cell surface changes in cold agglutination. Clin. Exp. Immun. 3:313, 1968.

89. Schatzmann, J. H.: Herzglyoside als Hemmstoffe fur den aktiven Kalium und Natriumtransport durch die Erythrozythenmembran. Helv. Physiol. Pharmacol. Acta 11:346, 1953.

90. Schrier, S. L., and Doak, L. S.: Studies of the metabolism of human erythrocyte membranes. J. Clin. Invest. 42:156, 1963.

91. —, Moore, L. D., and Chiapella, A. P.: Inhibition of human erythrocyte membrane mediated ATP synthesis by anti-D antibody. Amer. J. Med. Sci. 256:340, 1968.

92. Seaman, G. V. F., and Uhlenbruck, G.: Surface structure of erythrocytes from some animal sources. Arch. Biochem. 100:493, 1963.

93. Sears, D. A., Weed, R. I., and Swisher, S. N.: Differences in the mechanism of in vitro immune hemolysis related to antibody specificity. J. Clin. Invest. 43:975, 1964.

94. Shinagawa, Y., and Ogura, M.: Cholinesterase in erythrocyte membrane. Kagaku 31:554, 1961.

95. Shin, H. S., Snyderman, R., Friedmann, É., Mellors, A., and Mayer, M. M.: Chemotactic and anaphylatoxic, fragment cleaved from the fifth component of guinea pig complement. Science 162:361, 1968.

96. —, and Mayer, M. M.: The third component of the guinea pig complement system. II. Kinetic study of the reaction of EAC'4,2a with guinea pig C'3. Enzymatic nature of C'3 consumption, multiphasic character of fixation, and hemolytic titration of C'3. Biochemistry 3:57, 1966.

97. Sirchia, G., Ferrone, S., and Mercuriali, F.: The action of two sulfhydryl compounds on normal human red cells: relationship to red cells of paroxysmal nocturnal hemoglobinuria. Blood 25:502, 1965.

98. Sitomer, G., Stroud, R. M., and Mayer, M. M.: Reversible adsorption of C'2 by EAC'4: Role of Mg^{2+}, enumeration of competent SAC'4, two-step nature of C'2a fixation and estimation of its efficiency. Immunochemistry 3:57, 1966.

99. Skou, J. C.: The influence of some cations on the adenosine triphosphatase from peripheral nerves. Biochim. Biophys. Acta 23:394, 1957.

100. —: Further investigation on a Mg^{++} Na^+-activated adenosine triphosphatase, possibly related to the active linked transport of Na^+ and K^+ across the nerve membrane. Biochim. Biophys. Acta 42:6, 1960.

101. Smith, J. K., and Becker, E. L.: Serum complement and the enzymatic degradation of erythrocyte phospholipid. J. Immun. 100:459, 1968.

102. Solomon, A. K.: The permeability of the human erythrocyte to sodium and

potassium. J. Gen. Physiol. 36:57, 1952.

103. Springer, G. F.: Enzymatic- and non-enzymatic alterations of erythrocyte surface antigens. Bact. Rev. 27:191, 1963.

104. —, Nagai, Y., and Tegtmayer, H.: Isolation and properties of human blood group NN and meconium Vg antigens. Biochemistry 5:3254, 1966.

105. —: Mammalian erythrocyte receptors: Their nature and their significance in immunopathology. Bayer Symposium I, Vol. 47, 1969.

106. Stroud, R. M., Austen, K. F., and Mayer, M. M.: Catalysis of C'2 fixation by C'1a: reaction kinetics, competitive inhibition by TAMe and transferase hypothesis of the enzymatic action of C'1a on C'2, one of its natural substrates. Immunochemistry 2:219, 1965.

107. Tosteson, D. C., Cook, P., Andreoli, T., and Tieffenberg, M.: The effect of valinomycin on the ionic permeability of thin lipid membranes. J. Gen. Physiol. 50:2527, 1967.

108. —, and Hoffman, J. F.: Regulation of cell volume by active cation transport in high and low potassium sheep red cells. J. Gen. Physiol. 44:169, 1960.

109. —: Active transport, genetics and cellular evolution. Fed. Proc. 22:19, 1963.

110. —: Some properties of the plasma membranes of high potassium and low potassium sheep red cells. Ann. N. Y. Acad. Sci. 137:577, 1966.

111. Uhlenbruck, G., Wintzer G., and Wersdorfer, R.: Studien über den Aufbau der Zellmembran, insbesondere derjenigen des roten Blutkörperchens. Z. Klin. Chem. 5:281, 1967.

112. Whittam, R.: Potassium movements and ATP in human red cells. J. Physiol. (London) 140:479, 1958.

113. Winzler, R. J., Harris, E. D., Pekas, D. J., Johnson, C. A., and Weber, P.: Studies on glycopeptides released by trypsin from intact human erythrocytes. Biochemistry 6:2195, 1967.

114. —: A glycoprotein in human erythrocyte membranes. In Jamieson, G. A., and Greenwalt, T. J. (Eds.): Red Cell Membrane Structure and Function, Philadelphia, Lippincott, 1969, p. 157.

Mechanisms of Heinz Body Formation and Attachment to Red Cell Membrane

By Harry S. Jacob

A NUMBER OF MUTANT HEMOGLOBINS are inordinately unstable, denaturing in circulating red cells into Heinz bodies and precipitating when heated in vitro to 50°C.[7] Patients harboring these hemoglobins suffer from the syndrome of congenital Heinz body hemolytic anemia (CHBHA) which consists of: chronic hemolytic anemia; the presence of circulating red cells with inclusion (Heinz) bodies, more obvious after splenectomy; and frequently the excretion of urine darkened by accumulation of pyrrolic pigments, loosely termed "dipyrroles".[19,20,29] The disease is most often familial and is transmitted as an autosomal dominant trait; the affected patients harboring unstable hemoglobins such as Köln,[6] Zürich,[9] Genova,[28] and others. Sporadic cases involving hemoglobins such as Hammersmith[8] also occur and presumably represent spontaneous mutations in hemoglobin structure. Except for hemoglobin Zürich, most of the unstable hemoglobins produce a moderately severe, chronic hemolytic disease. Members of families with hemoglobin Zürich, in contrast, are healthy unless exposed to various oxidant drugs such as the sulfonamides or aminoquinoline antimalarials, at which times acute Heinz body hemolytic anemia ensues.[9] Regardless of their instability in vivo, the unstable hemoglobins are all precipitated by incubation at 50°C for brief periods, the basis of a simple screening test for CHBHA.[7] Splenectomy has been useful, although not totally curative, in most cases with debilitating anemia. In this regard it should be recognized that many of the unstable hemoglobins have increased oxygen affinity[2] resulting in tissue hypoxia beyond that predicted from hemoglobin levels only. Therefore, clinical parameters such as fatigue, tachycardia, poor growth, etc., become crucial in evaluating the necessity for splenectomy in these patients.

Recent studies from this and other laboratories have elucidated the biochemical mechanisms underlying the denaturation of the genetically unstable hemoglobins into Heinz bodies and the attachment of these inclusion bodies to the red cell membrane. In addition, such studies have also yielded insights into the causes of the hemolytic anemia and dipyrroluria of CHBHA. These studies will be reviewed.

Role of Heme Loss from Unstable Hemoglobins in Heinz Body Formation

In 1966, at a time when many of these studies were begun, the mutations

From University of Minnesota Medical School, Minneapolis, Minn.

Partially supported by USPHS Research Grants HE 10053, HE 12513 and HE 5600 from the National Institutes of Health, and by a Wellcome Research Travel Grant from the Wellcome Trust of London.

Harry S. Jacob, M.D.: Chief, Section of Hematology and Professor of Medicine, University of Minnesota Medical School, Minneapolis, Minn.

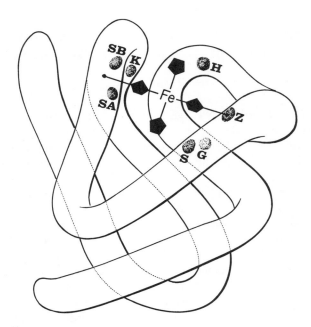

K = KÖLN; β^{98} Valine ⟶ Methionine

H = HAMMERSMITH; β^{42} Phenylalanine ⟶ Serine

Z = ZURICH; β^{63} Histidine ⟶ Arginine

G = GENOVA; β^{28} Leucine ⟶ Proline

S = SYDNEY; β^{67} Valine ⟶ Alanine

SA= SANTA ANA; β^{88} Leucine ⟶ Proline

SB= SABINE; β^{91} Leucine ⟶ Proline

Fig. 1.—Representative mutations in Heinz body-forming hemoglobins. The β chain of hemoglobin, after Perutz,[22] is depicted, emphasizing that the amino acid substitutions in several CHBHA hemoglobins closely neighbor the β chain heme group (portrayed as Fe-tetrapyrrole).

in only two unstable hemoglobins, Köln[6] and Zürich,[9] had been elucidated. We noted then that both involved substitutions at or near the sites where heme is bound to the β polypeptide chain of globin. We hypothesized that such mutations might alter the usual conformation of that area of the β chain which forms a close-fitting pocket in which the heme group resides; heme attachment might thereby be jeopardized. This concept seemed attractive because the postulated detached hemes could be catabolized to urinary "dipyrroles" explaining the prominent, dark pigmenturia of CHBHA. As described below, this hypothesis has been supported by evidence gathered in our laboratory with hemoglobins Köln and Hammersmith,[12,13] and its validity is further strengthened by observations of Bradley and his coworkers on another unstable hemoglobin, Gun Hill.[3] This hemoglobin causes moderately severe CHBHA and has been reported to have β chains entirely devoid of

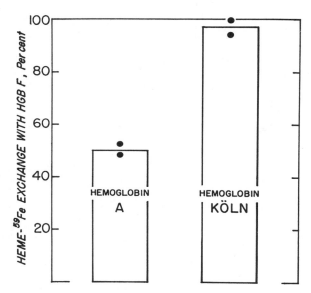

Fig. 2.—Diminished avidity of Hgb-Köln for heme. ^{59}Fe-ferrihemes from ^{59}Fe Hgb-Köln (right) fully exchange with originally unlabeled ferrihemes of Hgb-F in 90 minutes. Only about half of this number of ^{59}Fe hemes transferred from ^{59}Fe Hgb-A during the same time (left). (From Jacob, H. S. et al., courtesy Nature 218:1214, 1968.)

heme groups. A profound abnormality in the area of the heme pocket of the β chain exists in hemoglobin Gun Hill due to the deletion of the five amino acids adjacent to the proximal heme-binding histidine of the β chain (at β-92).

Since the original observations, several more unstable hemoglobins associated with CHBHA have been isolated and their mutations elucidated. Fig. 1 examines these mutations using the Perutz model[22] of β-chain structure. It is clear that nearly all the unstable hemoglobins reported to date have mutations in the vicinity of the β-chain heme group. This consistency supports the view that heme loss from β chains might frequently underlie the denaturation of CHBHA hemoglobins into Heinz bodies.

Direct evidence that the avidity of unstable hemoglobins for their heme groups is diminished was obtained in studies on hemoglobin Köln. Utilizing the observations of Bunn and Jandl[4] that ferrihemes are in constant flux between globin molecules, we found that this flux is excessive from hemoglobin Köln. As shown in Fig. 2, when unlabeled fetal methemoglobin is incubated with either methemoglobins Köln or A which have had their heme groups labeled with ^{59}Fe, the fetal hemoglobin becomes labeled. Thus, in 90 minutes ^{59}Fe-methemoglobin A exchanges approximately 50 per cent of its hemes with hemoglobin F. In contrast, at this time hemoglobin Köln has completely exchanged its hemes with hemoglobin F and may have done so even earlier.

More recently, excessive fluxes of heme from other unstable hemoglobins have been demonstrated and, in addition, actual heme depletion has been

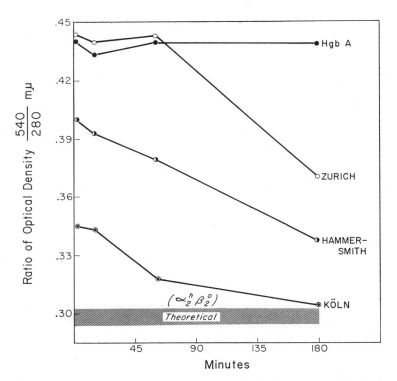

Fig. 3.—Heme loss from various oxyhemoglobins incubated at 50°C. Unlike Hgb-A (solid circles), Heinz body-forming hemoglobins progressively lose hemes as measured by decreasing 540/280 mμ ratios. Hemoglobins Köln and Hammersmith are somewhat heme-deficient prior to incubation and with heating develop 540/280 mμ ratios approaching that of a heme-protein completely devoid of β chain hemes, $\alpha_2^{\text{heme}}\beta_2^0$ (hatched bar).

documented in many CHBHA hemoglobins during their precipitation into Heinz bodies. When heated to 50°C these hemoglobins precipitate into a mass of typical coccoid Heinz bodies,[12,13] a phenomenon which underlies the simple screening procedure for the CHBHA syndrome described above. As shown in Fig. 3, heme loss occurs concomitantly with this instability. That is, the ratio of optical density at 540 mμ (mainly heme) to that at 280 mμ (mainly globin) diminishes in the heated CHBHA hemoglobins but remains stable in hemoglobin A (solid circles). The tendency of CHBHA hemoglobins to lose hemes correlates well with the severity of the clinical syndrome associated with them. Thus, hemoglobin Zürich loses hemes only after prolonged heating (open circles, Fig. 3), and in vivo causes no hemolytic anemia unless patients are stressed with certain oxidant drugs, such as the sulfonamides.[9] Conversely, hemoglobins Hammersmith and Köln produce a chronic, moderately debilitating Heinz body hemolytic anemia,[7,8] and in vitro immediately lose hemes. Indeed, like hemoglobin Gun Hill,[3] both are partially heme-depleted prior to incubation, simply following chromatographic isolation.

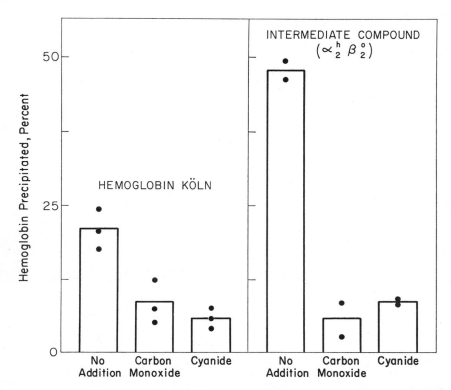

Fig. 4.—Inhibition of heat precipitation of Hgb-Köln (left) and $\alpha_2^{\text{heme}}\beta_2^{\,0}$ (right) by cyanide or carbon monoxide. Both heme-proteins copiously precipitate into Heinz bodies when heated at 50°C for two hours, unless the heme ligands, cyanide or carbon monoxide, are present.

Still another unstable hemoglobin with mutant β chains, hemoglobin San Francisco, causes moderate hemolysis in the patient and has also been shown to be heme-depleted after isolation.[17] During Heinz body generation at 50°C the CHBHA hemoglobins further diminished in 540/280 mμ ratio which ultimately approaches a value observed for a hemoglobin synthesized to contain no hemes at all on its β chains, $\alpha_2^{\text{heme}}\beta_2^{\,0}$, (hatched bar, Fig. 3).

This intermediary compound, which would form if hemes are lost preferentially from mutant β chains, has been synthesized and extensively studied by Winterhalter and his coworkers.[32-34] Together we have shown that $\alpha_2^{\text{heme}}\beta_2^{\,0}$ behaves identically to the genetically unstable hemoglobins in several ways; a mimicry which further supports the role of heme loss from β chains in the pathogenesis of Heinz body formation in CHBHA. Thus, $\alpha_2^{\text{heme}}\beta_2^{\,0}$ copiously precipitates into typical Heinz bodies at 50°C (left bar, right portion of Fig. 4). The heme ligands, cyanide or carbon monoxide, have been shown to suppress heme loss from hemoglobins,[4] a property which probably underlies their ability to inhibit the precipitation of heated unstable hemoglobins such as Köln (left portion, Fig. 4). As shown in the right portion

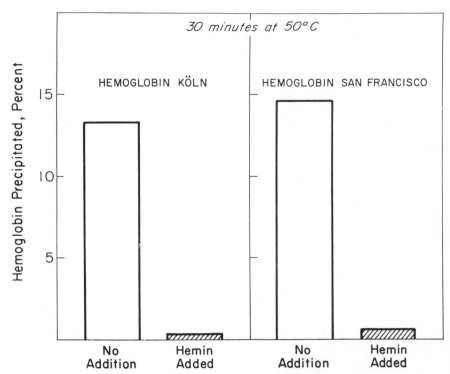

Fig. 5.—Inhibition of unstable hemoglobin precipitation by heme repletion. The CHBHA hemoglobins, Köln (left) and San Francisco (right), copiously precipitate at 50°C into Heinz bodies (open bars). Repletion of lost hemes by addition of crystalline hemin prior to incubation inhibits this denaturation (hatched bars).

of Fig. 4, these ligands also suppress Heinz body precipitation from heated $\alpha_2^{heme}\beta_2^{0}$. Conversely, oxidation of hemes to the ferric state accelerates their loss from globin polypeptide chains,[4] and approximately doubles Heinz body production from CHBHA hemoglobins,[12] and from $\alpha_2^{heme}\beta_2^{0}$.[18]

If $\alpha_2^{heme}\beta_2^{0}$ is supplemented with enough crystalline hemin to provide the normal stoichiometry of hemoglobin (four hemes per $\alpha_2\beta_2$ tetramer), physiologically normal hemoglobin A is produced.[33] This rejuvenated material no longer precipitates at 50°C. Similarly, if the genetically unstable hemoglobins, Köln and San Francisco, are treated with excess crystalline hemin prior to incubation at 50°C, their precipitation into Heinz bodies is virtually prevented (Fig. 5).

Finally, the observation that $\alpha_2^{heme}\beta_2^{0}$ migrates electrophoretically more slowly than heme-replete hemoglobin A at pH 8.6 (first channel, Fig. 6), suggested to us that heme deficiency might underlie the aberrant electrophoretic behavior of at least two unstable hemoglobins, Köln[6] and Sabine.[30] Both migrate more slowly than A, frequently in multiple bands, despite mutations which should lead to no net charge alterations. This paradox is shown in the second channel of Fig. 6 in which a split, electrophoretically

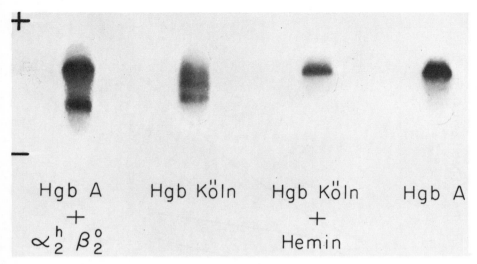

Fig. 6.—Effect of heme repletion on electrophoretic mobility of Hgb Köln. Starch gel patterns at pH 8.6 are depicted. Repletion of lost hemes of Hgb Köln by crystalline hemin (channel 3) changes the slow, multiple-banded molecule (channel 2) into a homogeneous one with mobility identical to Hgb A (channel 4). (From Jacob, H. S., and Winterhalter, K. H.; courtesy Proc. Nat. Acad. Sci. U.S.A. 65:697, 1970).

slow hemoglobin Köln is noted. Addition of excess crystalline hemin to synthetic $\alpha_2^{heme}\beta_2^o$ generates a hemoglobin both electrophoretically and physiologically identical to hemoglobin A.[33] Similarly, addition of hemin to hemoglobin Köln transforms it into a homogeneous hemeprotein with a mobility identical to hemoglobin A (third channel, Fig. 6).

Further work with synthetic $\alpha_2^{heme}\beta_2^o$ also has provided insight into the chemical nature of the Heinz body precipitate. When Heinz bodies are generated from $\alpha_2^{heme}\beta_2^o$ at 50°C, soluble free α^{heme} chains accumulate in the supernatant solution.[18] By implication, naked β^o chains must be the bulk of the precipitated Heinz body material. The whitish (e.g., heme-deficient) color of the precipitate strengthens this conjecture. Analogously, free α^{heme} chains progressively accumulate in supernates from heated solutions of several CHBHA hemoglobins. That precipitation of heme-deficient β chains into Heinz bodies probably also occurs in vivo in CHBHA is suggested by our observation and others[11] that soluble α^{heme} chains exist in fresh hemolysates from patients with CHBHA.

ALTERED SULFHYDRYL REACTIVITY OF UNSTABLE HEMOGLOBINS

Allen and Jandl,[1] in their studies of phenylhydrazine-induced denaturation of hemoglobin A into Heinz bodies, stressed the role of sulfhydryl metabolism in the oxidative destruction of hemoglobin. A sequence of oxidative damage was proposed as follows: (a) ferrohemes are initially oxidized to ferrihemes (methemoglobin); (b) if the oxidant stress continues, the two titratable

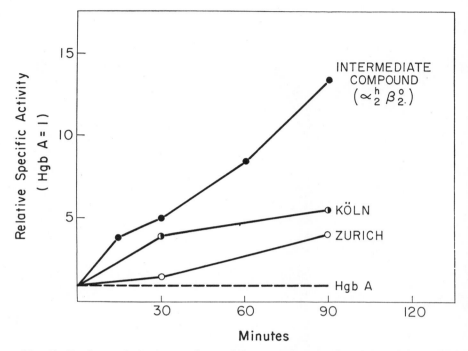

Fig. 7.—Binding of $G^{35}SH$ to hemoglobins. The unstable hemoglobins, Köln, Zürich, and especially synthetic $\alpha_2^{heme}\beta_2^{o}$, bind $G^{35}SH$ more readily than does Hgb-A (dotted line).

sulfhydryl groups of hemoglobin, present as cysteines at the 93rd position of the β chain, are then oxidized; (c) glutathione (GSH), present in large amounts in red blood cells, may bind to these oxidizing β-93 thiols in mixed disulfide linkage in what has been presumed to be a protective process; and (d) finally, further oxidation ultimately causes conformational changes in the α and β polypeptide chains in hemoglobin, exposing previously buried sulfhydryl groups, whose oxidation accompanies gross precipitation of the heme-protein into Heinz bodies.

We have gathered evidence that similar abnormalities in thiol homeostasis occur in denaturing CHBHA hemoglobins. Thus, the β-93 sulfhydryl groups in hemoglobin Köln and Hammersmith,[12,13] and perhaps in Ube-I as well,[31] are blockaded by mixed disulfide formation with GSH. The resulting diminution in unbound GSH levels, which has been noted in most cases of CHBHA,[7] stimulates hexose monophosphate shunt (HMP) activity, also a characteristic of these red cells.[7,12] This stimulation is understandable from the previous demonstration[16] that HMP shunt activity is inversely proportional to the concentration of free reduced glutathione within red cells; it therefore accelerates when GSH is bound to denaturing hemoglobins. In fact, as much as half of the total GSH of CHBHA red cells may be bound to the unstable hemoglobin.[12]

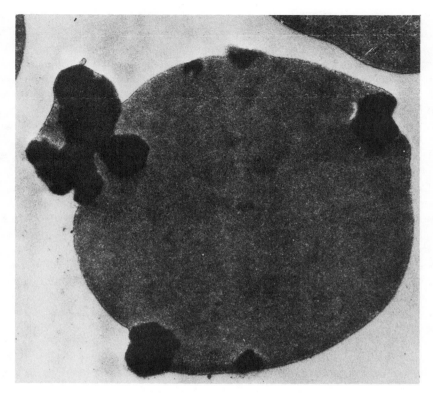

Fig. 8.—Electron microscopic view of Heinz bodies attaching to, and distorting, the membrane of a red blood cell. (From Rifkind, R. A., and Danon, D.; courtesy Blood 25:885, 1965.)

Since the free thiol group of hemoglobin (at β-93) neighbors the proximal heme-binding histidine of the β chain (at β-92), the reaction of the former with GSH might be expected to alter the conformation of the heme pocket; heme attachment could suffer thereby. In fact, it has been demonstrated that any manipulation of this sulfhydryl group, as with its blockade by agents such as N-ethylmaleimide or paramercuribenzoate, provokes heme detachment from the β chain.[4] As might be predicted, such treatment also potentiates Heinz body generation.[1,12] These observations intially led us to the view that all mutations in CHBHA act to enhance the oxidizability of the β-93 thiol; heme loss would be potentiated and ultimately denaturation of naked globin chains into Heinz bodies would occur. The data portrayed in Fig. 7 suggest, however, that heme loss from mutant beta chains in CHBHA hemoglobins is primary, and that excessive reactivity of β-93 thiols is secondary. That is, whereas the CHBHA hemoglobins, Zürich and Köln, bind $G^{35}SH$ in mixed disulfide linkage more readily than does hemoglobin A, this enhanced affinity is even more striking with $\alpha_2^{heme}\beta_2^{\circ}$ (solid circles, Fig. 7). These observations indicate that the hyperreactivity of the β-93 thiols of CHBHA hemoglobins probably reflect the lack of hemes on neighboring β-92 histidines. That $G^{35}SH$ in these studies indeed binds by mixed disulfide linkage to β-93 thiols has been demonstrated by ancillary studies. Thus, protein bound radioactivity is specifically released by disulfide-splitting reagents, and when $G^{35}SH$-labeled hemoglobins are cleaved into their component α and β polypeptide chains, only the latter contain label.[12] Since the β-93 thiol is the only titratable sulfhydryl group of intact hemoglobin,[10] it seems likely that it represents the $G^{35}SH$-binding group.

ATTACHMENT OF DENATURING HEMOGLOBINS TO RED CELL MEMBRANE THIOLS

During the early phases of denaturation of hemoglobin by phenylhydrazine, Heinz bodies are generated in the interior of red cells. The electron microscopic studies of Rifkind and Danon (Fig. 8) show that they later coalesce, migrate, and finally attach themselves to the red cell membrane.[27] We suspected that this attachment might involve binding by mixed disulfide linkage of the excessively reactive sulfhydryl groups of denaturing hemoglobins to the sulfhydryl groups of the red cell membrane; this in a manner analogous to the excessive binding of these hemoglobin thiols to cytoplasmic GSH. Several pieces of indirect evidence made this suspicion reasonable. Red cells with membrane thiols blockaded, as would occur in the proposed scheme, have been studied previously and mimic CHBHA red cells closely. The unique characteristic of these artifically sulfhydryl-inhibited cells is their rapid removal from the circulation, and their strikingly specific sequestration and destruction in the spleen.[15] CHBHA red cells behave similarly; that is, ^{51}Cr-red cell survivals in patients harboring hemoglobin Köln are about one-fifth normal, and body surface monitoring of radioactivity indicates the spleen to be the predominant site of cell sequestration and destruction.[12] The characteristics in vitro of red cells in which membrane thiols are blockaded also mirror those of CHBHA red cells. In both, cation leaks are accelerated, predisposing the cells to colloid osmotic hemolysis.[12,14] Similar findings have been reported in α-thalassemia red cells containing hemoglobin H inclusion bodies.[21]

Several experiments provided direct evidence that Heinz bodies are attached to red cell membrane thiols through mixed disulfide linkages. With phase optics, about 70 per cent of red cell ghosts from a recently studied splenectomized patient with hemoglobin Köln contained attached Heinz bodies. Treatment with mercaptoethanol, a potent disulfide-splitting reagent, reduced this number by half, whereas ethanol, a closely similar molecule but unreactive toward disulfide bonds, was without effect.[12] More evidence was obtained by utilizing measurements of binding of radiolabeled unstable hemoglobins to red cell ghosts. In the studies shown in Fig. 9, ^{59}Fe-labeled hemoglobins A or synthetic $\alpha_2^{heme}\beta_2^{\circ}$ of known specific activity were incubated with red cell ghosts. One aliquot of ghosts was untreated, the other had its membrane thiols preblockaded by exposure to paramercuribenzoate (PMB). Following removal of unreacted hemoglobin by washing, the radioactivity and hence the amount of heme-protein, bound to the ghosts could be assessed.

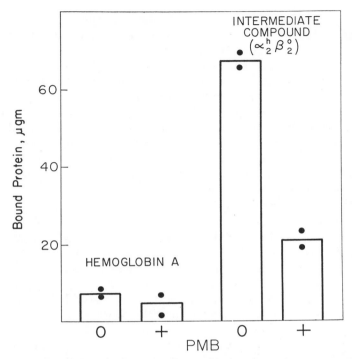

Fig. 9.—Diminished binding of hemoglobins to sulfhydryl-depleted (PMB-treated) red cell membranes. Preblockade of red cell ghost sulfhydryl groups with PMB inhibits their binding of ^{59}Fe-labeled $\alpha_2^{heme}\beta_2^{\circ}$ (right). By contrast, Hgb A binding is slight and negligibly affected by PMB (left).

Significantly more $\alpha_2^{heme}\beta_2^{\circ}$ was bound to normal red cell ghosts than hemoglobin A. The enhanced binding, however, was inhibited if thiols of the red cell membranes were blockaded by PMB prior to incubation with the labeled $\alpha_2^{heme}\beta_2^{\circ}$ (right portion, Fig. 9). Identical results have been published for hemoglobin Köln.[12] These observations support the involvement of membrane thiols in unstable hemoglobin attachment to red cell membranes. Conversely, the slight attachment of hemoglobin A to red cell ghosts probably was not by disulfide binding since preblockade of the membrane thiols with PMB was noninhibitory in this case (left portion, Fig. 9).

Finally, Heinz body-containing red cells have been found to by hypersusceptible to artificial membrane sulfhydryl blockade, consistent with their previous partial thiol depletion by reaction with denaturing hemoglobins. When treated with PMB at several different concentrations, hemoglobin Köln-containing red cells inordinately leak cations and become osmotically fragile as compared to identically treated normal red cells.[12]

Recent studies[26] have demonstrated that Heinz body-containing red cells become trapped in the spleen while traversing small apertures in basement membranes separating cords from sinusoids. These spaces are sufficiently small to require extreme deformation of red cells desiring passage through them. The rigid Heinz body acts as a "sticking point" and unless the remaining red cell can break away from its anchoring inclusion, its destruction is ordained. The presence of fragmented red cells within the circulation of patients with Heinz body hemolytic anemias reflects this sequence.[7] In addition, another, previously neglected, mechanism of red cell destruction in CHBHA is suggested by the studies shown in Fig. 9. Membrane thiols are probably blockaded by reaction with unstable hemoglobins even before they precipitate. As shown, *soluble* $\alpha_2^{heme}\beta_2^0$ attaches to membrane thiols through mixed disulfide bonds. By analogy to our previous studies of red cells with artifically blockaded membrane sulfhydryls,[14,15] this phenomenon should predispose CHBHA red cells to premature sequestration and osmotic destruction in the spleen even before the unstable hemoglobins have actually precipitated into Heinz bodies.

Fig. 10 summarizes current concepts of the pathogenesis of the CHBHA syndrome. It stresses the view that heme is crucial to globin stability and that its loss from mutant β chains is the crucial event in hemoglobin denaturation in CHBHA. As depicted, most of the congenital unstable hemoglobins contain mutations closely neighboring the β chain heme group. Heme binding suffers thereby and it seems likely that the excessive freed hemes are degraded to the pigmented dipyrroles so prominent in this syndrome. The resulting intermediary compound, lacking hemes on its β chains and analogous to our synthetic $\alpha_2^{heme}\beta_2^0$, is unstable, splitting into soluble heme-containing α chains (which have been observed in fresh hemolysates from patients with this syndrome) and into insoluble heme-depleted β chains which make up the bulk of the Heinz body material. These denaturing polypeptide chains, and soluble $\alpha_2^{heme}\beta_2^0$ as well, attach to cytoplasmic GSH and more importantly to red cell membrane thiols in mixed disulfide linkage. The resulting blockade of membrane thiols, in addition to the rigid inclusion body itself, predisposes the red cell to entrapment and destruction in reticuloendothelial organs, especially the spleen. It also causes increased red cell membrane permeability in a fashion analogous to that observed in other cells when proteins such as ADH[24] or insulin[5] bind membrane thiols in mixed disulfide linkage. In the latter instances, a useful function is served, whereas with the red cell, osmotic damage, premature reticuloendothelial entrapment, and hemolytic anemia result.

Although the above sequence of denaturation and hemolysis is thought to underlie most CHBHA hemoglobinopathies, new observations[25] on another, recently described unstable hemoglobin, Philly (β^{35} tyrosine→phenylalanine) suggest that a modification of this mechanism may be required. The mutation in this hemoglobin, and perhaps in hemoglobin Riverdale-Bronx[23] as well, is not near the heme pocket, but is in a region where α and β chains make contact. The linkage between these subunits evidently diminishes thereby, as free α and β chain monomers (with hemes attached) have been demonstrated during precipitation of this CHBHA hemoglobin.[25] Unlike most other CHBHA hemoglobins, hemoglobins Philly and Bronx-Riverdale produce Heinz bodies which are reddish (heme-replete) rather than white (heme-deficient), and affected

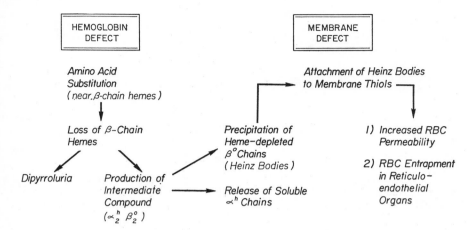

Fig. 10.—Scheme of the pathogenesis of CHBHA.

patients do not excrete dark "dipyrrolic" urine. These observations suggest that excessive monomer formation, even in the absence of heme loss, may be another, albeit less common, cause of denaturation of unstable hemoglobins into Heinz bodies.

ACKNOWLEDGMENT

Much of the original work presented was performed in collaboration with Doctors John V. Dacie and Michael Brain of the Royal Postgraduate Medical School of London, with Doctors Robin Carrell and Herman Lehmann of the University of Chambridge and with Dr. Kaspar H. Winterhalter of the Eidg. Technische Höchschule of Zürich, Switzerland. Their hospitality, friendship, and intellectual provocation are gratefully remembered.

REFERENCES

1. Allen, D. W., and Jandl, J. H.: Oxidative hemolysis and precipitation of hemoglobin., II. Role of thiols in oxidant drug action. J. Clin. Invest. 40:454, 1961.

2. Bellingham, A. J., and Huehns, E. R.: Compensation in haemolytic anaemias caused by abnormal haemoglobins. Nature 218:924, 1968.

3. Bradley, T. B., Jr., Wohl, R. C., and Rieder, R. F.: Hemoglobin Gun Hill: deletion of five amino acid residues and impaired heme-globin binding. Science 157:1581, 1967.

4. Bunn, H. F., and Jandl, J. H.: Exchange of heme among hemoglobin molecules. Proc. Nat. Acad. Sci. USA 56:974, 1966.

5. Carlin, H., and Hechter, O.: The disulfide-sulfhydryl interchange as a mechanism of insulin action. J. Biol. Chem. 237:1371, 1962.

6. Carrell, R. W., Lehmann, H., and Hutchison, H. E.: Hemoglobin Köln (β-98 valine→methionine): an unstable protein causing inclusion-body anaemia. Nature 215:915, 1966.

7. Dacie, J. V., Grimes, A. J., Meisler, A., Steingold, L., Hemsted, E. H., Beaven, G. H., and White, J. C.: Hereditary Heinz-body anaemia. A report of studies on five patients with mild anaemia. Brit. J. Haemat. 10:388, 1964.

8. —, Shinton, N. K., Gaffney, P. J., Jr., Carrell, R. W., and Lehmann, H.: Haemoglobin Hammersmith (β42 (CDI) Phe→ Ser.) Nature 216:663, 1967.

9. Frick, P. G., Hitzig, W. H., and Betke, K.: Hemoglobin Zürich. I. A new hemoglobin anomaly associated with acute hemolytic episodes with inclusion bodies after sulfonamide therapy. Blood 20:261, 1962.

10. Guidotti, G., and Konigsberg, W.:

The characterization of modified human hemoglobin. I. Reaction with iodoacetamide and n-ethylmaleimide. J. Biol. Chem. 239:1474, 1964.

11. Huehns, E. R., and Schooter, E. M.: Human hemoglobins. J. Med. Genet. 2:48, 1965.

12. Jacob, H. S., Brain, M. C., and Dacie, J. V.: Altered sulfhydryl reactivity of hemoglobins and red blood cell membranes in congenital Heinz body hemolytic anemia. J. Clin. Invest. 47:2664, 1968.

13. —, —, —, Carrell, R. W., and Lehmann, H.: Abnormal haem binding and globin SH group blockade in unstable hemoglobins. Nature 218:1214, 1968.

14. —, and Jandl, J. H.: Effects of sulfhydryl inhibition on red blood cells. I. Mechanism of hemolysis. J. Clin. Invest. 41:779, 1962.

15. —, and —: Effects of sulfhydryl inhibition on red blood cells. II. Studies in vivo. J. Clin. Invest. 41:1514, 1962.

16. —, and —: Effects of sulfhydryl inhibition on red blood cells. III. Glutathione in the regulation of the hexose monophosphate pathway. J. Biol. Chem. 241:4243, 1966.

17. —, and Winterhalter, K. H.: Unstable hemoglobins: the role of heme loss in Heinz body formation. Proc. Nat. Acad. Sci. U.S.A. 65:697, 1970.

18. —, and —: The role of hemoglobin heme loss in Heinz body formation: studies with a partially heme-deficient hemoglobin and with genetically unstable hemoglobins. J. Clin. Invest. (in press).

19. Lange, R. E., and Akeroyd, J. H.: Congenital hemolytic anemia with abnormal pigment metabolism and red cell inclusion bodies: a new clinical syndrome. Blood 13:950, 1958.

20. Miwa, S., Kato, H., Saito, M., Cheba, A., Irisawa, K., and Ohyama, H.: Congenital hemolytic anemia with abnormal pigment metabolism and red cell inclusion bodies. Report of a case and review of literature. Nippon Ketsueki Gakkai Zasshi 28:593, 1965.

21. Nathan, D. G., and Gunn, R. B.: Thalassemia: the consequences of unbalanced hemoglobin synthesis. Amer. J. Med. 41:815, 1966.

22. Perutz, M. F.: Structure and function of haemoglobin. I. A tentative atomic model of horse oxyhaemoglobin. J. Mol. Biol. 13:646, 1965.

23. Ranney, H., Jacobs, A. S., Udem, L. and Zalusky, R.: Hemoglobin Riverdale-Bronx. An unstable hemoglobin resulting from the substitution of arginine for glycine at helical residue B6 of the β polypeptide chain. Biochem. Biophys. Res. Commun. 33:1004, 1968.

24. Rasmussen, H., Schwartz, I. L., Schoessler, M., and Hochster, G.: Studies on the mechanism of action of vasopressin. Proc. Nat. Acad. Sci. USA 46:1278, 1960.

25. Rieder, R. F., Oski, F. A., and Clegg, J. B.: Hemoglobin Philly (β35 Tyr→Phe): Clinical and chemical consequences of the loss of a pair of inter-chain hydrogen bonds. Clin. Res. 17:340, 1969.

26. Rifkind, R. A.: Heinz body anemia: an ultrastructural study. II. Red cell sequestration and destruction. Blood 26:433, 1965.

27. —, and Danon, D.: Heinz body anemia: an ultrastructural study. I. Heinz body formation. Blood 25:885, 1965.

28. Sansone, G., Carrell, R. W., and Lehmann, H.: Haemoglobin Genova: β28 (B10) Leucine→Proline. Nature 214:877, 1967.

29. Schmid, R., Brecher, G., and Clemens, T.: Familial hemolytic anemia with erythrocyte inclusion bodies and a defect in pigment metabolism. Blood 14:991, 1959.

30. Schneider, R. G., Ueda, S., Alperin, J. B., Brimhall, B., and Jones, R. T.: Hemoglobin Sabine beta 91 (F7) Leu→Pro. An unstable variant causing severe anemia with inclusion bodies. New Eng. J. Med. 280:739, 1969.

31. Shibata, S., Iuchi, I., Miyaji, T., Ueda, S., and Takeda, I.: Hemolytic disease associated with the production of abnormal hemoglobin and intraerythrocytic Heinz bodies. Nippon Ketsueki Gakkai Zasshi 26:164, 1963.

32. Winterhalter, K. H.: Sequence of linkage between the prosthetic groups and the polypeptide chains of haemoglobin. Nature 211:932, 1966.

33. —, Amiconi, G., and Antonini, E.: Functional properties of a hemoglobin-carrying heme only on α chains. Biochemistry 7:2228, 1968.

34. —, and Deranleau, D. A.: The structure of a hemoglobin carrying only two hemes. Biochemistry 6:3136, 1967.

Alteration of Membrane Deformability in Hemolytic Anemias

By P. L. LaCelle

ERYTHROCYTES must bend and deform into diverse shapes as they pass through the capillary circulation which at many points has a diameter considerably less than the erythrocyte itself. From the time of Rous,[67] many investigators[5,7,8,10,29,30,38,47,51,73] have observed the marked degree of deformation required and the capacity of the erythrocyte to return to the normal biconcave shape when stress is removed.

Considerable new understanding concerning the physical properties of the erythrocyte, and some of the relationships of the erythrocyte and the circulatory system, have been provided to the clinician by the research technics and insights of biophysicist and engineer, as well as those in a variety of other disciplines. Burton's monograph[11] is a lucid introduction to the biophysical considerations, and some basic physical concepts pertinent to erythrocytes and rheology have been outlined by Reiner[63] and Wayland.[82]

Judging from the degree of deformation in the capillaries and the observations in vitro,[61] the erythrocyte is deformed by relatively small forces. Hence, it might be predicted that the physical properties of the cell are extremely important in permitting the requisite flexibility. Further, it might be postulated that small changes in mechanical properties of the cell can render it less able to conform to the microcirculation and thus prejudice survival of such an altered erythrocyte, particularly in passage through the spleen where the severest restriction (approximately 3 μ mean diameter[83]) to red cell passage may exist.[13]

It is evident that the cell shape, the physical properties of the membrane and the state of the cell contents all may contribute to the deformability or flexibility of the erythrocyte. The biconcave disc shape of the normal erythrocyte results in an advantageous surface area-to-volume relationship. Provided that the membrane is deformable and the cell contents liquid, the erythrocyte can be expected to adapt its volume configuration to conform to the specific requirements of the microcirculation. A sphere enclosing the

From the University of Rochester School of Medicine and Dentistry, Rochester, N.Y.

Supported by USPHS Research Grant HE-06241-08 and the United States Atomic Energy Project at the University of Rochester and has been assigned publication number UR-49-1181.

P. L. LaCelle, M.D.: Associate Professor of Medicine and of Radiation Biology and Biophysics, University of Rochester School of Medicine and Dentistry, Rochester, N.Y.

maximum volume per area is a geometrically rigid body and it is predictable that a spherical cell is poorly adapted to survive in the microcirculation. Any factor causing spherocytosis will diminish deformability and, hence, survival.

The erythrocyte membrane can withstand great bending forces, but probably less than a 15 per cent increase in area, without rupture and hemolysis. The membrane is thought in ideal terms to possess elastic and viscous properties and in some cases a plastic element,[88] i.e., a constituent which, if distorted sufficiently, will not reversibly regain its initial dimensions and physical character. Factors modifying the membrane itself, independent of shape, may be expected to alter its flexibility by changing these properties.

Cell contents in the normal state may be assumed to be liquid[22,27,31,61] and the fluidity would not be expected to limit the cellular deformability. Obviously, in those situations where cell contents are altered, e.g., sickle hemoglobin in circumstances of low oxygen tension, the cell contents may be significant determinants of cell rigidity.

The mechanical properties of erythrocytes have potential importance in understanding possible mechanisms of cellular destruction. They are altered in all hemolytic states, suggesting that they are intimately involved in determining erythrocyte life span. Alteration of cellular deformability may be estimated by viscometry of cell suspensions, in which case the derived data are a function of the contributions of cell shape, membrane deformability and viscosity of the contents of the cell populations. Similarly, the filterability of cells, measured as the capacity to pass or the rate of passage through a calibrated filter system, estimates the shape and intrinsic membrane properties determining deformability. Contribution of the cell shape and membrane to the flexibility of the cell also may be measured in terms of forces required to deform individual erythrocytes in a glass micropipette. This method has the advantage of determining properties of single cells, the capacity to distinguish between shape and membrane contribution to cell deformability and is a convenient method for examining the properties of erythrocyte ghosts in which the internal environment has been manipulated to assess the effect of various chemical compounds on cell shape and membrane deformability.

It is the purpose of this review to describe technics which have been employed to measure erythrocyte deformability, to indicate pathologic states in which abnormal deformability has been demonstrated, and to outline some experimental observations made in an attempt to define some of the factors which may cause changes in deformability. Particular emphasis will be given to micropipette studies since the micropipette (actually a glass microcapillary) provides the simplest simulation of a capillary and permits evaluation of individual erythrocyte behavior by direct observation under conditions of controlled pressure (driving force) and rate of deformation.

ABNORMAL VISCOSITY OF ERYTHROCYTE SUSPENSIONS IN PATHOLOGIC STATES

Viscometric technics useful for measurement of cell deformability employ viscometers capable of indicating net viscosity of cell suspensions over a wide range of shear rates. A variety of instruments exists, ranging from the cone-plate viscometer (e.g., Brookfield Model LVT) and the modifications of the

Table 1.—Specific Measurements of Erythrocyte Deformability by Viscometry, Filtration Methods and Indirect Methods

Viscometry of Cell Suspensions
Charache and Conley 1964,[15] 1967[16]
Dintenfass 1964,[21] 1968,[25] 1968[26]
Braasch and Hennig 1965,[5] Braasch and Jennet 1968[6]
Chien et al. 1967,[17] 1969[19]
Ham et al. 1968[39,40]
Wells et al. 1968[89]
Schmid-Schönbein et al. 1968,[69,70] 1969[71,72]
Nevaril et al. 1968[58]
Usami 1969[81]
Weed et al. 1969[86]

Filtration of Cell Suspensions
Jandl et al. 1961[45]
Prothero and Burton 1962[60]
Bergentz and Danon 1965[3]
Ultmann and Gordon 1965[80]
Charache et al. 1967[16]
Gregersen et al. 1967,[36] 1968[37]
Ham et al. 1968[39,40]
Murphy 1968[55]
Bygdeman and Wells 1968,[12] Wells 1968[89]
Chien et al. 1969[19]

Radial Migration of Erythrocytes
Goldsmith and Mason 1969[35]

Couette viscometer (e.g., Gilinson-Dauwalter-Merrill,[33] an instrument permitting measurements of extremely low shear rates) to the unique instruments such as the rotational ring-in-ring or cone-in-cone and parallel plate viscometers utilized by Dintenfass.[20,23,24] Schmid-Schönbein and Wells[70] have employed a direct visualization cone-plate microviscometer in which it is possible to observe and photograph the rheologic behavior of cell suspensions. The simple capillary type viscometers, e.g., Ostwald, give data at relatively high shear rates. Table 1 indicates the apparatus used by several investigators interested in erythrocyte viscometry.

In principle, viscometers measure the resistance of layers of cells to slide over each other at different velocities. At high shear rates, i.e., rapid application of force to cause relatively high velocities of cells moving past each other, the system in vitro is analogous to larger blood vessels where force application rate and velocity are similar and cells move in rouleaux. Such a system is relatively limited in its usefulness in understanding the contribution of cell shape and membrane properties to deformability, especially at low hematocrits (i.e., normal physiologic range). At hematocrits above 60 per cent, cells must be deformable because no closer packing without deformation is possible, as is readily seen from Burton's calculations.[11] The sensitivity of viscometry as a measure of deformability increases at high hematocrit values, a practical optimum being 80 per cent. With such great packing, the necessity for close cell contact obviously increases and the necessity to deform to pass each other is more critical.[11,86] At lower shear rates, typical of the microcirculation where cells move independently with bolus flow (plasma interspersed between cells in single file), viscometric data may reflect with greater sensitivity the cell parameters crucial to cell passage in 3–12 μ diameter capillaries.[86] More meaningful data are obtained when carefully washed cells are used, to obviate the contribution of plasma proteins, especially fibrinogen, to the viscosity. The pH must be controlled, for, as Dintenfass has shown,[28] decreased pH increases observed viscosity.

Several investigators have studied the viscosity of cells containing abnormal hemoglobins and have measured the increased viscosity apparent in hereditary spherocytosis and in acquired hemolytic disease. These findings are listed in Table 2.

Viscometry was used by Weed et al. to evaluate the metabolic factors altering erythrocyte deformability.[86] During incubation without substrate, a decrease in erythrocyte ATP, with attendant increase of membrane calcium content, led to a marked increase in viscosity at a time when the disc-to-sphere transformation occurred. Restoration of cellular ATP by incubation of depleted cells with adenosine restored normal viscosity. A yield stress was found (a positive shear stress intercept at zero shear in Casson plot) in fresh cells at a hematocrit of 80 per cent and the yield stress was observed to increase four-fold in ATP-depleted erythrocytes. Hemoglobin from ATP-depleted cells showed no alteration of viscosity. The shear dependence and yield stress of reversible viscosity changes after ATP depletion were consistent with a sol-gel change in hemoglobin or non-hemoglobin protein. Since these findings were also characteristic of white, hemoglobin-free ghosts, it would

Table 2.—Compilation of Hemolytic Disorders in Which Decreased Erythrocyte Deformability Has Been Documented*

Disorder	Technic Employed	Reference
Hb-SS	Millipore filtration	Jandl et al. 1961[42]
HS†		
HS-SS	Falling ball viscometry	Charache and Conley 1964[15]
Hb-SA (D,J)		
Hb-SS	Cone-in-cone viscometry	Dintenfass 1964[21]
Hb-SA		
Hb-CC	Paper filtration	Teitel 1964[75]
	Millipore filtration	Charache et al. 1967[16]
	Cone plate viscometry	
HS	Millipore filtration	Murphy 1967[54]
	Cone plate viscometry	
Hb-CC	Millipore filtration	Murphy 1968[55]
	Cone plate viscometry	
Hemolytic anemia	Cinephotomicrography	Bond et al. 1968[4]
HbSS	Rheoscopy	Schmid-Schönbein and Wells 1969[72]
HS	Micropipette	LaCelle and Weed 1969[49]

*Modified from Teitel[75]
†Hereditary spherocytosis.

appear likely that membrane non-hemoglobin protein, as well as residual hemoglobin, may have been responsible for the altered viscosity.

ERYTHROCYTE FILTERABILITY

The filters employed for the study of erythrocyte deformability have consisted of three general types:

Paperfiber filters. These inexpensive filters have long multibranching channels of dimensions approximately equal to the erythrocyte diameter (e.g., Schleicher & Schuell, No. 589) and a thickness many times greater than the cell diameter. Their effectiveness as a means whereby to estimate deformability lies in the fact that erythrocytes pack the tortuous channels of the filter such that there are narrow "effective" channels through which cells may deform to pass under the force of gravity, i.e., the hydrostatic pressure of the column of cells above the filter. Thus, if the same type of paper is employed for control cells and pathologic or altered cells in suspension, a comparison may be made of deformability. Weed has shown that a good correlation exists between the information derived from simple filterpaper filteration and that obtained by the cone plate (Brookfield LVT) viscometer or the Gilison-Dauwalter-Merrill modified Couette[33] viscometer on suspensions with high hematocrits at low shear rates.[86] Teitel et al. have employed paper filters to examine cells from a wide variety of pathologic states[75-78] and have documented the dependence of normal cellular deformability on reduced membrane sulfhydryl groups and hemoglobin integrity.[77,78] Both Teitel[76] and Weed et al.[86] have shown that filterability decreases as erythrocyte ATP is depleted.

Millipore* filters. This membrane filter consists of an inert matrix penetrated by tubular channels having considerably greater uniformity of diameter and virtually no branching when compared to paper filters. Because such filters are not easily wetted, wetting agents (namely detergents) have been added; such compounds cause variable lysis.[86] Jandl et al.[44,45] in early studies utilizing such filters of 5 μ pore diameter had observed that sickle cells at low pO$_2$ are selectively retained by the filter at pressures as high as 120 mm. Hg (approximately 4× end-arteriolar) pressure. Cells agglutinated by complete antibody can be forced through the filter by pressures greater than 4 mm. Hg, whereas cells coated with incomplete agglutinin are retained by the filter only at low pressures in the presence of fibrinogen or another rouleaux-inducing agent. In patients with hereditary spherocytosis, the spherical cells are trapped by the filter. This early work had demonstrated the potential value of calibrated filters in the study of erythrocyte abnormalities.

Polycarbonate† sieves. These filters consist of polycarbonate membranes of less than 15-μ thickness through which right cylindrical channels of very uniform dimensions have been produced by X-irradiation. The ratio of pore area to total membrane area is small and pore density, diameter and length

---

*Millipore Corporation, Bedford, Mass.
†Nucleopore Membrane Filters, General Electric Co., Pleasanton, Calif.

may be documented by electron microscopy. Gregerson et. al. have found that the "critical pore diameter," i.e., 3.0-μ pore filters through which 100 per cent of normal erythrocytes pass, corresponds to the theoretical calculation based on the assumption that the erythrocyte membrane is deformable but not distensible.[36] Erythrocytes, hardened in acetaldehyde, have a higher viscosity than normal cells and demonstrate the importance of cell deformation in passing through small pores; such cells cannot pass a 6.8 μ pore.

Deformability of Individual Cells and Factors Affecting Deformability

Rand, with a micropipette technic, has documented several pertinent facts concerning the erythrocyte membrane.[61,62] The normal erythrocyte membrane can withstand large bending forces without irreversible damage, but tangential stress, i.e., force tending to increase the area, leads to a marked increase in the membrane tension with a dramatic increase in stiffness or resistance to deformation. "Stretch" or increase of area by more than 10 to 15 per cent leads to hemolysis. In measuring membrane deformability by the micropipette technic, similar values are noted at all parts of the membrane, including the rim, suggesting that within the limits of sensitivity of the method, membrane stiffness and, therefore, composition is probably quite uniform over the entire cell surface. Tensions in the membranes of cells swollen in hypotonic medium are high, as would be predicted from the geometrically rigid spherical shape. Rand has also been able to document the viscoelastic nature of the erythrocyte membrane.[62]

In experiments employing the micropipette technic, we have attempted to separate the contribution of membrane resistance to bending or deformation from that of cell shape in the deformability of the cell. To accomplish this goal, we have measured the force required to induce a uniform hemispherical bulge in the tip of a calibrated pipette (Diagram A in Fig. 1). Assuming that this change occurs with no stretching of membrane or change in volume, this deformation closely approaches an isochoric applicable deformation[31] and no change in membrane stresses should occur. The force required to deform the membrane in this fashion should be a function of the membrane's mechanical properties resulting from its composition and does not include the shape contribution because the normal biconcave disc has excess area with respect to volume enclosed, and the membrane is slack, i.e., tension is minimal, while the measurement is being made. This force has been designated P, to distinguish it from force P_t, a measure of pressure required to pull the cell into and through the micropipette. It is evident that the deformation to a cylindrical shape to conform to the micropipette will involve total cell deformability, including the contribution made by shape, and as the cell becomes spherical its resistance to passage through the cylindrical channel, analogous to the capillary of the microcirculation, will increase. The rigid glass pipette parallels the relative inelasticity of capillaries. Burton[11] has noted that, despite their simple composition, capillaries distend less than 0.08 per cent per mm. Hg pressure, and cannot be burst by as much as 500 mm. Hg.

Canham and Burton[13] have described the "minimum cylindrical diameter"

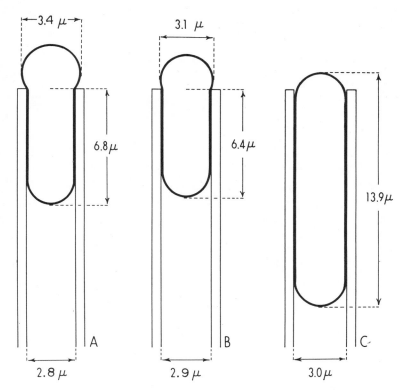

Fig. 1.—Deformation of fresh normal erythrocytes in pipettes of various diameters. (Cell dimensions are means from 50 cells.) Reprinted by permission from Red Cell Membrane Structure and Function, Jamieson, G. A., and Greenwalt, T. J. (Eds.). Philadelphia, J. B. Lippincott, 1969.

as defining the limiting cylindrical volume configuration permitting passage all cells to pass, whereas 2.85 μ permits the cell to advance in the pipette configuration by a "critical radius" and "critical pore diameter", respectively. The diameter, 3.66 μ according to Canham and Burton's calculations, would permit the cell to pass without damage; Gregersen et al. calculate 2.84 μ. In our studies (Fig. 1), a 2.75 μ nominal aperture* is not sufficient to permit all cells to pass, whereas 2.85 μ permits the cell to advance in the pipette to the point where a greater than 3-μ diameter sphere remains outside the pipette. As P_t is increased, the cell passes through the pipette without hemolysis. In Fig. 1C, one notes that the cell in a 3.0 μ pipette has deformed to a length of 13.9 μ from an original diameter of 8.3 μ and thus the cell does not change volume or area significantly ($V = 91 \ \mu^3$, $A = 144.6 \ \mu^2$). To obtain a length of 13.9 μ, some "linear stretch" without net area change or hemolysis must have occurred. It is of interest that our "critical diameter" with the micropipette is also approximately 3 μ, as predicted by observations of

---

*Measurements made with an eyepiece micrometer calibrated with a stage micrometer.

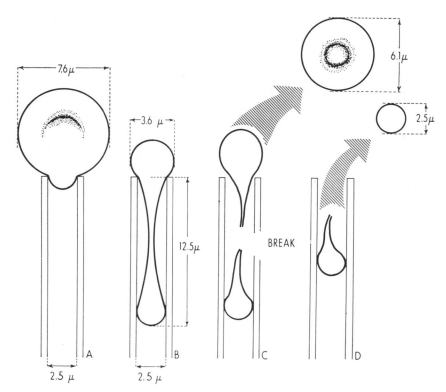

Fig. 2.—Deformation and fragmentation of fresh normal erythrocytes in 2.45 μ diameter pipette. (A) Hemispherical deformation at $-P = 4$ mm. H_2O. (B,C). Increase of $-P$ leads to rupture of the membrane with resulting microsphere and smaller biconcave disc. (Dimensions are means from 50 cells.)

other investigators. Burton[10,13] has suggested that the "minimum cylindrical diameter" in vivo probably is found in the spleen in the light of its increase after splenectomy, a suggestion readily appreciated by clinicians who observe improvement of hemolytic disorders after splenectomy. Actually, the critical diameter may be postulated to be less than 3 μ in the spleen because the thickness of the basement membrane separating splenic cords from sinuses is less than 13.9 μ, and thus the channel determined by the aperture diameter and the basement membrane thickness is less restrictive than that of the 3 μ diameter by 14 μ length channel implied in the micropipette study or calculated by Gregersen. Following splenectomy, the minimum diameter within the general circulation increases to approximately 4 μ.[85]

FRAGMENTATION

Fragmentation is recognized[64,66,83,84] as a mechanism for determining, in part, erythrocyte life span in normal cells and particularly in Heinz body anemias, as well as in microangiopathic hemolytic anemia.[9] Figure 2 illustrates a fragmentation sequence observed with a small micropipette with a resulting microsphere and a biconcave disc of reduced size. No detectable hemolysis occurs in Fig. 2C; after fragmentation, the ruptured membrane reseals. As many as three or four fragments may be produced by such maneuvers resulting, finally, in a rigid sphere as membrane area decreases. Fragmentation in vivo presumably occurs as portions of cells "stick" to epithelial cells, rough surfaces, or areas of fibrin mesh[9] with tearing of fragments from the cell. The net result is similar, production of spherical rigid cells. Such a mechanism may be contributory to the eventual destruction of normal cells by inducing trapping in the reticuloendothelial system, particularly in the spleen where 1 to 2 per cent of blood passes through the critical apertures of the cord system[44] and clearly leads to fragmentation in Heinz body disorders.[90] Splenic pH may be as low as 6.8,[54] a factor which increases membrane rigidity[48] and thus enhances fragmentation and trapping of rigid cells.

METABOLIC FACTORS INFLUENCING CELLULAR AND MEMBRANE DEFORMABILITY

Nakao demonstrated that ATP is essential to the maintenance of the biconcave disc shape of the normal erythrocyte[57] and to posttransfusion survival,[56] an observation thus directly linking metabolic integrity of the cell to its survival. To define the relationship of ATP, cell shape and membrane properties essential to survival in vivo, Weed et al.[86] have attempted to correlate changes in the ATP-depleted state with altered membrane properties:

It was found that in normal cells undergoing substrate depletion, initial change in membrane deformability was observed at 4 to 6 hours when cellular ATP was approximately 70 per cent of the initial value. After ATP decreased at 10 hours to 15 per cent of the initial value, progressive parallel increases in membrane calcium and membrane stiffness occurred. After 24 hours of depletion, a point where no membrane lipid was lost, membrane calcium was 400 per cent and membrane stiffness (pressure P in the micropipette method) 1000 per cent greater than values in fresh cells. In ATP-depleted cells, residual hemoglobin and non-hemoglobin proteins increased 200 and 60 per cent respectively. Regeneration of ATP in depleted cells by incubation in adenosine produced significant reversal in all these parameters, and both shape and deformability returned to normal.

Introduction of calcium into ghosts made from fresh erythrocytes transformed them immediately into rigid spheres; magnesium and barium did not have this effect. Indeed, magnesium simultaneously incorporated in tenfold excess almost completely prevented the calcium effect (Table 3). ATP or EDTA in tenfold molar excess completely blocked the effect of calcium in the fresh cell ghosts. When introduced into rigid ghosts made from depleted cells, ATP or EDTA restored deformability of depleted cell ghosts to normal fresh cell ghost values. Incorporation of adenosine diphosphate, nicotinamide adenine dinucleotide, NADH, NADPH, glutathione, inosine triphosphate, guanosine triphosphate, or uridine triphosphate into depleted cell ghosts had no effect on the membrane stiffness or spherical shape. Introduction of ouabain during depletion and during ATP regeneration had no

Table 3.—Membrane Deformability of Red Cell Ghosts

	P	SE	N
	mm.H$_2$0		
Reconstituted Ghosts Prepared From Fresh Cells			
Control, no additive	4.2	0.10	108
+10^{-4} M Ca	352.0	10.00	96
+10^{-4} M Ca and 10^{-3} M ATP	4.3	0.15	29
+10^{-4} M Ca and 10^{-3} M EDTA	4.7	0.27	29
+10^{-4} M Ca and 10^{-3} M Mg	4.2	0.19	50
+ 1 mM NADP	6.8	0.52	11
+ 1 mM NAD	6.2	0.38	11
Reconstituted Ghosts Prepared From 24-hour Depleted Cells			
Control, no additive	51.2	1.21	58
+10^{-3} M ATP	4.5	0.12	109
+10^{-3} M EDTA	4.1	0.15	30
+10^{-3} M Mg	9.5	1.10	12
+ 1 mM NADPH	58.5	4.20	7
+ 1 mM NADH	64.5	2.89	10
+10^{-3} M UTP	49.2	2.99	5
+10^{-3} M GTP	50.1	1.92	12
+10^{-3} M ITP	66.6	6.00	5
+10^{-3} M ADP	53.6	4.71	10
+ 3 × 10^{-3} M ADP	58.7	3.30	9
+ 5 × 10^{-3} M GSH	47.8	1.61	29
Hemoglobin-free Ghosts			
Prepared from fresh cells	3.8	0.10	50
Prepared from 24-hour depleted cells	24.3	1.05	30
Prepared from fresh cells, incubated 24 hours + 10^{-4} M Ca	33.3	0.55	50
Prepared from fresh cells, incubated 24 hours + 10^{-4} M Ca + 10^{-3} M Mg	5.6	0.36	50

ATP, adenosine triphosphate; EDTA, ethylenediamine tetra-acetate; NAD nicotinamide adenine dinucleotide; NADP, NAD phosphate; NADPH, reduced NADP; UTP, uridine triphosphate: GTP, guanosine triphosphate; ITP, inosine triphosphate; ADP, adenosine diphosphate; GSH, reduced glutathione.

effect, suggesting that the ouabain-inhibitable sodium-potassium pump was not intimately involved in the determination of cell shape and deformability.

These studies emphasize the important direct effect of ATP in maintenance of the marked deformability characteristics or maximal elastic compliance of the membrane, as well as to maintain the advantageous biconcave disc shape of the cell. The locus of the ATP function is intracellular, at the inner membrane surface, since ATP incorporated into depleted cell ghosts is efficacious, whereas external ATP is not. It is presumably a calcium chelating effect, duplicated by EDTA but not by other organic phosphates, and magnesium may exert a similar effect by competing with calcium for a specific membrane site. Small local changes in membrane ATP/Ca relationships in the erythrocyte membrane may lead to changed membrane deformability before total cell ATP and calcium values have changed markedly. In the light of the increase in the hemoglobin and nonhemoglobin protein accumulation in the depleted cell, it is postulated that a subcortical sol-to-gel transformation occurs as local membrane calcium rises, with the net result of a thickened membrane, stiffened by the accumulated protein gel. Dintenfass considers that such a mechanism at such an interface would increase viscosity.[22] The existence of actomyosin-like proteins in the membrane, suggested by the work of Marchesi,[52] the ATP-dependent shape changes,[57] and Wins and Schoffeniels' calcium and ATP-dependent contraction of ghosts,[91] as well as the demonstration by Rosenthal[65] of calcium-dependent ATPase activity in fibrillar protein from erythrocytes, would provide a second calcium-dependent mechanism for increasing membrane stiffness, i.e., contraction of the membrane to a more dense, less compliant structure.

In studies of erythrocytes stored in ACD solution, disc-to-sphere transformation, fragmentation loss of lipids and a striking increase in viscosity and decrease in filterability were observed by Haradin et al.[41] Post-transfusion survival was correlated with ATP-dependent parameters of shape, filterability and viscosity, and improved survival after incubation in adenosine was not associated with reversal of lipid loss or critical hemolytic volume. These authors suggested that ACD solution was a particularly good one in which to store cells because the citrate in the medium would chelate and reduce the amount of calcium which might accumulate in the erythrocytes. In a study of cell deformability[48] utilizing the micropipette technic, it was evident that mean deformability of stored cells correlated with reported post-transfusion survival of ACD stored blood. At three weeks of storage, only 76 per cent of stored cells could traverse a 2.85 μ pipette, one through which all fresh cells could pass. P_t increased linearly with storage time. Similar changes were observed in ghosts, identifying the process as one localized to the membrane; lower pH caused a twofold increase in P_t. Due to shape change and intrinsic membrane stiffness, the stored cells required a larger "minimum critical diameter" than normal cells and had a greater tendency to be trapped mechanically.

OXYGEN TENSION AND DEFORMABILITY

Chanutin,[14] Benesch and Benesch,[2] Akerblom et al.,[1] Garby[32] and subsequently others have described the relationship between the deoxygenation of hemoglobin and the increased affinity of deoxygenated hemoglobin for organic phosphates, particularly 2,3-diphosphoglycerate and ATP. Tosteson[79] has noted increased ion movements in deoxygenated cells. In recent studies,[50] we have observed that whereas the erythrocyte membrane retains its extremely deformable character down to po_2 values of 30 mm. Hg (normal venous po_2 being 38–42 mm. Hg), further decrease results in marked stiffness of the erythrocyte, reminiscent of the rigidity observed in ghosts into which calcium has been introduced (Table 3). The significance of this finding is readily appreciated if one considers that po_2 values are as low as 25 mm. Hg in the coronary sinus, 15–30 mm. Hg in the maternal side of the placenta,

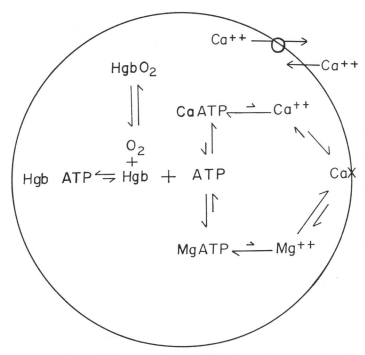

Fig. 3.—Postulated model for regulation of erythrocyte membrane deformability, involving interactions of ATP, calcium, magnesium, and membrane protein. A possible mechanism for effective ATP "depletion" by deoxyhemoglobin-ATP binding is indicated. Details of the model are described in the text.

and also presumably as low in portions of the venous circulation where there is pathologic stasis. Furthermore, if the po_2 of the splenic cord system is low, further rigidity would be induced in erythrocytes in this particularly hazardous portion of the circulation.

We have interpreted the rigidity associated with lowered oxygen tension to be a consequence of hemoglobin binding of ATP with resultant diminution of the ATP concentration at the inner membrane surface and concomitant calcium-induced stiffening and contraction (Fig. 3).

SULFHYDRYL GROUPS AND DEFORMABILITY

Membrane SH⁻ groups, essential membrane protein ligands well characterized by a variety of SH⁻ reacting compounds, are known to be functionally related to glucose transport and cation permeability. Altered membrane permeability is associated with decreased survival, and Jacob and Jandl[43] have observed that splenic sequestration of PCMB (parachloromercuribenzoate) damaged erythrocytes is not necessarily preceded by sphering. In micropipette experiments utilizing PCMB and inorganic mercury, we have noted a significant increase in P_t in the absence of sphering after two hours of incubation when the PCMB concentration in the incubation medium is 5×10^{-4} M and that of mercury is 10^{-5} M. Thus, altered deformability, prior to significant change in electrolyte and water content, may be a significant determinant of splenic trapping and subsequent destruction of erythrocytes.

MEMBRANE PLASTICITY

In experiments with spherical tumor cells, Weiss has demonstrated the existence of a plastic element in the membrane, i.e., an element which has a yield point in terms of magnitude and rate of force application beyond which it will not resume its original configuration.[88] In examining normal white cells, particularly lymphocytes, we have observed a similar tendency to retain the deformed shape. In cells from patients with thalassemia trait, a tendency to retain the distorted shape is evident and does not appear to be related to the rate of application of the deforming pressure. Poikilocytes from such blood, when examined at various points on their periphery, also present differing values of P_t, suggesting local variations in the membrane flexibility of these cells. It might be postulated that such cells may have altered membrane properties due to excess globin chain accumulation without visible Heinz bodies and/or that such cells' intrinsic membrane composition, differing from normal cells, cannot tolerate the stresses in the circulation without distortion or rearrangement of structure to yield regions of relative stiffness. Abnormal plasticity has been observed in ovalocytosis as well.

MEMBRANE DEFORMABILITY IN HEREDITARY SPHEROCYTOSIS (HS)

The hereditary spherocyte is less deformable than the normal erythrocyte, both in terms of intrinsic membrane deformability and in total cellular deformability. P is greater than the normal cell mean in 50 per cent of fresh HS cells from eleven patients, and P_t in 95 per cent of HS cells is greater than the normal cell mean. After 24 hours of incubation without glucose, HS cells accumulate more calcium than normal cells and require greater quantities of ATP to prevent the deleterious effects of simultaneously introduced calcium in fresh HS ghosts. It is known from the work of Hoffman[42] that calcium accumulation can increase cation permeability, probably due to altered calcium binding to the membrane. Sodium flux is increased in HS, as it is in other hemolytic anemias. Incubation of fresh HS cells in adenosine to enhance membrane associated ATP lowers sodium influx to rates observed in normal cells. We interpret these data to indicate that the HS membrane is qualitatively similar to normal, but that it has a quantitatively different ATP requirement to prevent adverse calcium interaction at the inner membrane surface. It is postulated that HS cells differ in their membrane affinity for calcium and that increased calcium-membrane interaction leads to decreased deformability and increased cation fluxes. Whether or not the abnormal calcium affinity is the result of abnormal HS membrane proteins has not yet been demonstrated.

SUMMARY

In disorders which include metabolic depletion that may be typical of normal cell aging and stored erythrocytes, unbalanced hemoglobin synthesis, and cells modified by sulfhydryl inhibitors, viscosity measurements, filtration of erythrocyte suspensions and evaluation of force required to deform erythrocytes in micropipettes indicate abnormal erythrocyte deformability. In each disorder, loss of intrinsic membrane deformability, or alteration of shape due to loss of membrane by fragmentation or secondary to metabolic depletion and/or increased rigidity of cellular contents, leads to rigidity of the cell and predisposition to destruction in the spleen where factors of pH and po_2 may increase membrane stiffness. Alteration of deformability appears to be a common mechanism of destruction in hemolytic anemias.

REFERENCES

1. Akerblom, O., de Verdier, C.-H., Garby, L., and Högman, C.: Restoration of defective oxygen-transport function of stored red blood cells by addition of inosine. Scand. J. Clin. Lab. Invest. 21:245, 1968.

2. Benesch, R., and Benesch, R. E.: Ogygen and ion transport in red cells. Science 160:83, 1968.

3. Bergentz, S. E., and Danton, D.: Alterations in red blood cells of traumatized rabbits. I. Decreased filterability. Acta Chir. Sand. 130:165, 1965.

4. Bond, T. P., Guest, M. M., and Derrick, J. R.: High speed cinephotomicrography of the microcirculation in patients with hemolytic anemia. Europ. Soc. Microcirculation, International Conference, Gothenburg, (Sweden), 1968, Abstracts, p. 24.

5. Braasch, D., and Hennig, W.: Erythrocytenflexibilität and Strömungswiderstand in Capillaren mit einem Durchmesser unter 20μ Pflügers Arch. 286:76, 1965.

6. —, and Jennet, W.: Micro-methods for determination of blood flow resistance in small capillaries in vitro. Europ.' Soc. Microcirculation, International Conference, Gothenburg, Sweden, 1968, Abstracts, p. 172.

7. Branemark, P.-I., and Lindström, J.: Shape of circulating blood corpuscles. Biorheology 1:139, 1963.

8. —, Aspegren, K., and Breine, U.: Microcirculatory studies in man by high resolution vital microscopy. Angiology 15:329, 1964.

9. Bull, B. S., Rubenberg, M. D., Dacie, J, V., and Brain, M. C.: Microangiopathic haemolytic anaemia: Mechanisms of red-cell fragmentation: in vitro studies. Brit. J. Haemat. 14:643, 1968.

10. Burton, A. C.: Role of geometry, of size and shape, in the microcirculation. Fed. Proc. 25:1753, 1966.

11. —: Physiology and Biophysics of the Circulation. Chicago, Yearbook Medical Publishers, 1965, p. 50.

12. Bygdemann, S., and Wells, R.: Comparison of rheologic effects of isoviscous concentrations of dextran 40 and 150 upon erythrocyte aggregation deformability and cell viscosity. Europ. Soc. Microcirculation, International Conference, Gothenburg, Sweden, 1968, Abstracts, p. 30.

13. Canham, P. B., and Burton, A. C.: Distribution of size and shape in populations of normal human red cells. Circ. Res. 22:405, 1968.

14. Chanutin, A., and Curnish, R. R.: Effect of organic and inorganic phosphates on the oxygen equilibrium of human erythrocytes. Arch. Biochem. 121:96, 1967.

15. Charache, S., and Conley, C. L.: Rate of sickling of red cells during deoxygenation of blood from persons with various sickling disorders. Blood 24:25, 1964.

16. —, —, Waugh, D. F., Ugoretz, R. J., and Spurrell, J. R.: Pathogenesis of hemolytic anemia in homozygous hemoglobin C disease. J. Clin. Invest. 46:1795, 1967.

17. Chien, S., Usami, S., Taylor, H. M., Lundberg, J. L., and Gregersen, M. I.: Effects of hematocrits and plasma protein on human blood rheology at low shear rates. J. Appl. Physiol. 21:81, 1966.

18. Chien, S., Usami, S., Dellenback, R. J., and Gregersen, M. I.: Blood viscosity: influence of erythrocyte deformation. Science 157:827, 1967.

19. —, —, —, Bryant, C. A., and Gregersen, M. I.: Change of erythrocyte deformability during fixation in acetaldehyde. Internat. Soc. Hemorheology, Second International Conference, Heidelberg, Germany, 1969, Abstracts, p. 31.

20. Dintenfass, L.: An application of a cone-in-cone viscometer to the study of

viscometry, thixotropy and clotting of blood. Biorheology 1:91, 1963.

21.—: Rheology of packed red blood cells containing hemoglobins A-A, S-A, and S-S. J. Lab. Clin. Med. 64:594, 1964.

22. —: Molecular and rheological considerations of the red cell membrane in view of the internal fluidity of the red cell. Acta Haemat. 32:299, 1964.

23. —: A ring-in-ring adapter for the cone-in-cone rotational viscometer. Biorheology 2:221, 1965.

24. —: Viscosity of blood at high haematocrits measured in microcapillary (parallel-plate) viscometers of r=3-30 microns. Biorheology, Proc. 1st International Conference on Haemorheology, Reykjavik, 1966.

25. —: Fluidity (internal viscosity) of the erythrocytes and its role in physiology and pathology of circulation. Haematologia (Budapest) 2:19, 1968.

26. —: Fluidity (internal viscosity) of the red cell and the critical capillary radius and their role in physiology and pathology of the microcirculation. Europ. Soc. Microcirculation, International Conference, Gothenburg, Sweden, 1968, Abstracts, p. 52.

27. —: Internal viscosity of the red cell and a blood viscosity equation. Nature 219:956, 1968.

28. —, and E. Burnard: Effect of hydrogen ion concentration on in vitro viscosity of packed red cells and blood at high hematocrits. Med. J. Aust. 1:1072, 1966.

29. Fåhraeus, R., and Lindquist, T.: The viscosity of the blood in narrow capillary tubes. Amer. J. Physiol. 96:562, 1931.

30. Fitz-Gerald, J. M.: Mechanics of red-cell motion through very narrow capillaries. Proc. Roy Soc. (Biol.) 174:193, 1969.

31. Fung, Y. C.: Theoretical considerations of the elasticity of red cells and small blood vessels. Fed. Proc. 25:1761, 1966.

32. Garby, L., Gerber, G., and de Verdier, C.-H.: Binding of ATP and 2,3-DPG to hemoglobin. In Deutsch, E., Gerlach, E., and Moser, K. (Eds.): Metabolism and Membrane Permeability of Erythrocytes and Thrombocytes. Stuttgart, Georg Thieme, 1968, p. 66.

33. Gilinson, P., Jr., Daulwalter, C. R., and Merrill, E. W.: A rotational viscometer using an A.C. torque-to-balance loop and air bearing. Trans. Soc. Rheol. 7:319, 1963.

34. Ginn, F. L., Hochstein, P., and Trump, B. F.: Membrane alterations in hemolysis: internalization of plasmalemma induced by primaquine. Science 164:843, 1969.

35. Goldsmith, H. L., and Mason, S. G.: Some model experiments in hemodynamics. Internat. Soc. Hemorheology, Second International Conference, Heidelberg, Germany, 1969, Abstracts, p. 43.

36. Gregersen, M. I., Bryant, C. A., Hammerle, W. E., Usami, S., and Chien, S.: Flow characteristics of human erythrocytes through polycarbonate sieves. Science 157:825, 1967.

37. —, —, Chien, S., Dellenback, R. J., Magazinovic, V., and Usami, S.: Species differences in flexibility and deformation of erythrocytes (RBC). Europ. Soc. Microcirculation, International Conference, Gothenburg, Sweden, 1968, Abstracts, p. 67.

38. Guest, M. M., Bond, T. P., Cooper, R. G., and Derrick, J. R.: Red blood cells: change in shape in capillaries. Science 142: 1319, 1963.

39. Ham, T. H., Dunn, R. F., Sayre, R. W., and Murphy, J. R.: Physical properties of red cells as related to effects in vivo. I. Increased rigidity of erythrocytes as measured by viscosity of cells altered by chemical fixation, sickling and hypertonicity. Blood 32:847, 1968.

40. —, Sayre, R. W., Dunn, R. F., and Murphy, J. R.: Physical properties of red cells as related to effects in vivo. II. Effect of thermal treatment on rigidity of red cells, stroma, and the sickle cell. Blood 32:862, 1968.

41. Haradin, A. R., Weed, R. I., and Reed, C. F.: Changes in physical properties of stored erythrocytes: relation to survival in vivo. Transfusion 9:229, 1969.

42. Hoffman, J. F.: Cation transport and structure of the red cell plasma membrane. Circulation 26:1201, 1962.

43. Jacob, H. S., and Jandl, J. H.: Effects of sulfhydryl inhibition in red blood cells. II. Studies in vivo. J. Clin. Invest. 41:1514, 1962.

44. —: Sequestration of reticulocytes and of abnormal red cells by filtration at low pressures. J. Clin. Invest. 37:905, 1958.

45. —, Simmons, R. L., and Castle, W. B.: Red cell filtration and the pathogenesis of certain hemolytic anemias. Blood 18:133, 1961.

46. —, and Aster, R. H.: Increased splenic pooling and the pathogenesis of hyper-

splenism. Amer. J. Med. Sci. 253:383, 1967.

47. Krogh, A.: The Anatomy and Physiology of Capillaries. New Haven, Yale University Press, 1922, 1929.

48. LaCelle, P. L.: Alteration of deformability of the erythrocyte membrane in stored blood. Transfusion 9:238, 1969.

49. —, and Weed, R. I.: Abnormal membrane deformability. A model for the hereditary spherocyte. J. Clin. Invest. 48: 48a, 1969.

50. —: The effect of oxygen tension on erythrocyte deformability. Submitted for publication.

51. Macfarlane, R. G., and Robb-Smith, A. H. T.: Functions of the Blood. New York, Academic Press, 1961.

52. Marchesi, V. T., and Steers, E., Jr.: Selective solubilization of a protein component of the red cell membrane. Science 159:203, 1968.

53. Merrill, E. W., Gilliland, E. R., Cokelet, G., Shin, H., Britten, A., and Wells, R. E., Jr.: Rheology of human blood, near and at zero flow. Effects of temperature and hematocrit level. Biophys. J. 3:199, 1963.

54. Murphy, J. R.: The influence of pH and temperature on some physical properties of normal erythrocytes and erythrocytes from patients with hereditary spherocytosis. J. Lab. Clin. Med. 69:758, 1967.

55. —: Hemoglobin CC disease: rheological properties of erythrocytes and abnormalities in cell water. J. Clin. Invest. 47:1483, 1968.

56. Nakao, K., Wada, T., and Kamiyama, T.: A direct relationship between adenosine triphosphate-level and in vivo viability of erythrocytes. Nature 194:877, 1962.

57. Nakao, M., Nakao, T., and Yamazoe, S.: Adenosine triphosphate and maintenance of shape of the human red cell. Nature 187:945, 1960.

58. Nevaril, C. G., Lynch, E. C., Alfrey, C. P., and Hellums, J. D.: Erythrocyte damage and destruction induced by shearing stress. J. Lab. Clin. Med. 71:784, 1968.

59. Prothero, J. W., and Burton, A. C.: The physics of blood flow in capillaries. II. The capillary resistance to flow. Biophys. J. 2:199, 1962.

60. — and —: The physics of blood flow in capillaries. III. The pressure required to deform erythrocytes in acid-citrate-dextrose, Biophys. J. 2:213, 1962.

61. Rand, R. P., and Burton, A. C.: Mechanical properties of the red cell membrane. I. Membrane stiffness and intracellular pressure. Biophys. J. 4:115, 1964.

62. —: Mechanical properties of the red cell membrane. II Viscoelastic breakdown of the membrane. Biophys. J. 4:303, 1964.

63. Reiner, M.: Deformation, Strain and Flow. New York, Interscience, 2nd ed., 1960.

64. Rifkind, R.: Destruction of injured red cells in vivo. Amer. J. Med. 41:711, 1966.

65. Rosenthal, A. S., Kregenow, E. M., and Moses, H. L.: Some characteristics of a Ca^{2+}-dependent ATPase activity associated with a group of erythrocyte membrane proteins which form fibrils. Biochim. Biophys. Acta 196:254, 1970.

66. Rous, P., and Robertson, O. H.: The normal fate of erythrocytes. I. The findings in healthy animals. J. Exp. Med. 25:651, 1917.

67. Rous, P.: Destruction of the red blood corpuscles in health and disease. Physiol. Rev. 3:75, 1923.

68. Schmid-Schönbein, H., Gaehtgens, P., and Hirsch, H.: Eine neue Methode zur untersuchung, der rheologischen Eigenschaften von Erythrocyten-Aggregaten. Pflügers Arch. 297:107, 1967.

69. —, —, and —: On the shear rate dependence of red cell aggregation in vitro. J. Clin. Invest. 47:1447, 1968.

70. —, Wells, R., and Schildkraut, R.: Demonstration of microscopic visualization of erythrocyte aggregation as a function of rate of shear in a Wells-Brookfield transparent cone-plate viscometer (Rheoscope). Europ. Soc. Microcirculation, International Conference, Gothenburg, Sweden, 1968, Abstracts, p. 175.

71. —, Goldstone, J., and Wells, R.: Model experiment in red cell rheology: the mammalian erythrocyte as a fluid drop. Internat. Soc. Hemorheology, Second International Conference, Heidelberg, Germany, 1969, Abstracts, p. 30.

72. —, and Wells, R.: Red cell deformation and red cell aggregation: Their influence on blood rheology in health and disease. Internat. Soc. Hemorheology, Second International Conference, Heidelberg, Germany, 1969, Abstracts, p. 25.

73. Skalak, R., and Branemark, P. I.: Deformation of red blood cells in capillaries.

Science 164:717, 1969.

74. Taylor, H. M., Chien, S., and Gregersen, M. I.: Comparison of viscometric behavior of suspensions of polystyrene latex and human blood cells. Nature 207:77, 1965.

75. Teitel, P.: Microrheological competence of erythrocytes and pathogenic mechanisms in hemolytic anemias. Proc. Symposium on Nuclear Medicine, Bucharest, October 22–24, 1969.

76. —: Disk-sphere transformation and plasticity alteration of red blood cells. Nature 206:409, 1965.

77. —: A haemorheological view on molecular interactions between red blood cell constituents in the pathogenesis of constitutional haemolytic anaemias. Internat. Soc. Hemorheology, Second International Conference, Heidelberg, Germany, 1969, Abstracts, p. 30.

78. —, Marcu, I., and Xenakis, A.: Erythrocyte microrheology: its dependence on the reduced sulfhydryl groups and hemoglobin integrity. Folia Haemat. (Leipzig) 90:281, 1968.

79. Tosteson, D. C., and Robertson, J. S.: Potassium transport in duck red cells. J. Cell. Physiol. 47:147, 1956.

80. Ultmann, J. E., and Gordon, C. S.: The removal of in vitro damaged erythrocytes from the circulation of normal and splenectomized rats. Blood. 26:49, 1965.

81. Usami, S., Chien, S., and Gregersen, M. I.: Viscometric behavior of young and aged erythrocytes. Internat. Soc. Hemorheology, Second International Conference, Heidelberg, Germany, 1969, Abstracts, p. 32.

82. Wayland, H.: Rheology and microcirculation. Bibl. Anat. 5:3, 1965.

83. Weed, R. I., and Weiss, L.: The relationship of red cell fragmentation, occurring within the spleen to cell destruction. Trans. Ass. Amer. Physicians 79:426, 1966.

84. —, and Reed, C. F.: Membrane alterations leading to red cell destruction. Amer. J. Med. 41:681, 1966.

85. —: The cell membrane in hemolytic disorders. Plenary Session Papers, XII. Cong. Internat. Soc. Hemat., New York, 1968, p. 81.

86. —, LaCelle, P. L., and Merrill, E. W.: Metabolic dependence of red cell deformability. J. Clin. Invest. 48:795, 1969.

87. Weiss, L.: The structure of fine splenic arterial vessels in relation to hemoconcentration and red cell destruction. Amer. J. Anat. 111:131, 1962.

88. —, and Clement, K.: Studies on cell deformability: some rheological considerations. Exp. Cell. Res. 58:379, 1969.

89. Wells, R., Bygdeman, S., and Schmid-Schönbein, H.: Analysis of viscous deformation of the red cell and its effect upon microvascular flow. Europ. Soc. Microcirculation, International Conference, Gothenburg, Sweden. 1968, Abstracts, p. 159.

90. Wennberg, E., and Weiss, L.: The structure of the spleen and hemolysis. Ann. Rev. Med. 20:29, 1969.

91. Wins, P., and Schofleniels, E. ATP+Ca++-linked contraction of red cell ghosts. Arch. Int. Physiol. 74:812, 1966.

Anatomical Hazards to the Passage of Erythrocytes Through the Spleen

By Leon Weiss and Mehdi Tavassoli

P ASSAGE THROUGH THE SPLEEN constitutes a hazard for blood-borne cells which may result in their modification or death. Certain hazards are more pronounced in the spleen than elsewhere. Thus, if normal spleen is perfused by normal and spherocytic red cells, normal cells pass through, while spherocytes are retained.[6] Erythrocytes moderately damaged by incomplete antibody or heat circulate through the remaining organs of the body but are trapped and destroyed in the spleen.[19] Foà-Kurloff cells, distinctive inclusion-containing lymphocytes, are selectively removed from the circulation by the spleen. A large number of granulocytes damaged by endotoxin are removed in the spleen.[21] The site of destruction of normal red cells at the end of their life span has been considered to be the marrow, but the case for this is not definitive and it may well be that they are destroyed in the spleen.[20] After splenectomy, the life-span of the erythrocytes is not measurably changed, but a significant change in life-span of red cells may go undetected by the methods presently available. Following splenectomy, defective erythrocytes marked by nuclear and cytoplasmic residua are present in the circulation.[20] In animals with a spleen, these marred cells are absent. The modifications in cells during splenic passage include the removal or pitting of hemosiderin from siderocytes[4] and the removal of inclusions from phenylhydrazine-damaged erythrocytes.[9,15] These cells may thereby be restored to a normal or near normal state. But, since they lose plasma membrane along with the inclusion, they approach the status of spherocytes. Being less plastic, they are more easily destroyed by deformation or other mechanical test. Immunologically competent cells may be selectively removed from the circulation by the spleen, and undergo transformation to plasma cells engaged in antibody production. Monocytes may be trapped from the blood stream and be induced to undergo transformation into macrophages.

The spleen, then, is a large filtration bed, sifting about 2 liters of blood per minute in the human adult.[6] It is the purpose of this review to present the anatomy of the organ and to consider how the structure contributes to the distinctive hazards imposed by the spleen upon cells in passage.

Anatomical Considerations

The spleen has a thick capsule and a rich trabecular system which by

From the Department of Anatomy, Johns Hopkins University School of Medicine, Baltimore, Md.

Leon Weiss, M.D.: *Professor of Anatomy, Department of Anatomy, The Johns Hopkins University School of Medicine, Baltimore, Md.* Mehdi Tavassoli, M.D.: *Research Fellow Department of Anatomy, The Johns Hopkins University School of Medicine, Baltimore, Md.; Blood Research Laboratory, Tufts-New England Medical Center, Medford, Mass.*

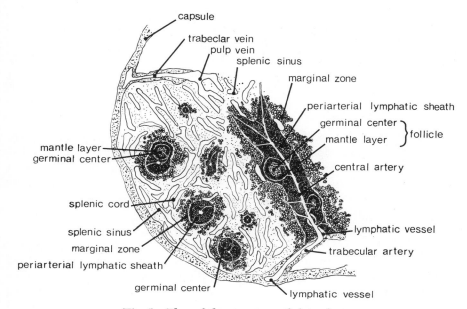

Fig. 1.—Plan of the structure of the spleen.

second, third and fourth degree branchings compartmentalizes the organ and carries vessels and nerves deep into the parenchyma (Fig. 1).

Arterial vessels travel through the trabeculae and leave them to enter the pulp. On entering the pulp, the artery is surrounded by a sleeve of loose reticular connective tissue whose interstices are packed with small lymphocytes and other free cells. This sleeve is termed the *periarterial lymphatic sheath*. The artery travels in the central axis, coaxially with the sheath and is termed the *central artery*. It sends branches of a minor sort into the sheath but most of the branches go to the periphery of the sheath or beyond it and end there. Here and there the sheath accommodates an ovoid accumulation of lymphocytes which may be very large. These are *follicles*, recognizable grossly as *Malpighian corpuscles*. They may contain *germinal centers*. When a germinal center is present the small lymphocytes of the follicle which form a shell about it are recognized as a mantle zone. Periarterial lymphatic sheath and follicles comprise the *white pulp*.

The periarterial lymphatic sheath is dependent upon the thymus. It is likely that its small lymphocytes are produced and released from the thymus and are selectively concentrated in the loose connective tissue surrounding the central artery. Thymectomy is followed by depletion of the lymphocytes from this zone. This type of depletion occurs also as a result of thoracic duct cannulation which diverts the small lymphocytes of thoracic duct lymph (largely of thymus origin)[14] from entrance into the blood. These lymphocytes are immunologically competent and are associated with delayed hypersensitivity and antigen recognition.[14] The small lymphocytes of the periarterial lymphatic sheath appear to follow a distinctive pathway. They travel in the blood through branches of the central artery and reach large postcapillary venules, the splenic sinuses, just outside the sheath in a location termed the *marginal zone* (see below). The small lymphocyte crosses the wall of the sinus and enters the sheath.[7] It then probably crosses the sheath centripetally, moving toward the central artery. Efferent lymphatic vessels are present in the trabeculae and often entwine the central arery. These vessels collect small lymphocytes and carry them out of the spleen and through the thoracic duct back to the blood.

The follicles and germinal centers are sites of antibody production. The germinal centers participate in primary responses[16] but are particularly active in anamnestic responses.[23] They are not dependent upon the thymus. In avian spleens they are likely to be dependent upon the bursa of Fabricius; the mammalian counterpart of the bursa has not been discovered.

The periarterial lymphatic sheaths are surrounded by a broad but variable zone recognized as the *marginal zone*. The marginal zone consists of a fine meshed reticular connective tissue which receives perhaps the greatest concentration of arterial endings. These are arterial vessels which spring from the central artery and its branches and run through the periarterial lymphatic sheath and into the marginal zone. Often, these vessels curve back toward white pulp just before ending. The marginal zone also contains a great number of venous sinuses, many of which abut upon the periarterial lymphatic sheaths. The marginal zone is continuous with the cords of the red pulp (see below). The interstices of the reticular meshwork of the marginal zone are jammed with free cells. Macrophages, lymphocytes, plasma cells and erythrocytes are typically abundant among these cells.

The salient function of the marginal zone is filtration of blood. This is accomplished by a large number of arterial vessels emptying into a fine-meshed three-dimensional web. This filtration is evident by the accumulation in the marginal zone of particulate matter, cells and other blood-borne material. The filtration is enchanced by the narrowing of the interstices of the meshwork by the free cells held in it. Not only do these cells render the filter finer by their physical bulk, but many of these cells are macrophages and plasma cells that may phagocytize material and produce antibody, thus further impeding passage.

As the central artery goes on it becomes finer, and the periarterial lymphatic sheath becomes thinner and terminates. The artery now becomes the *artery of the red pulp*. It may bifurcate into very slender noncommunicating branches, the *penicilli*, and soon after this, the vessels terminate. The red pulp consists, principally, of *cords* and *vascular sinuses*. The vascular sinuses are thin-walled, cucumber-shaped venous vessels which anastomose, forming the first elements in the blood efferent system. These vessels are also present in the marginal zone. They drain into pulp veins and thence into trabecular veins. From there the blood finds its way out of the organ. Sinuses are well developed in certain spleens of which the human spleen is an example. The endothelium consists of tapered rods which lie with their long axis parallel with the long axis of the vessel. The endothelial cells simply abut one another; no specialized membranous attachments have been observed. The endothelium lies upon a basement membrane distinctive for being so defective that it is reduced to strands, regularly organized, of basement membrane material which is composed of mucopolysaccharide. Outside the basement membrane the sinus is clothed by an incomplete layer of *adventitial cells* which may be phagocytic and which send processes into the contiguous cords. In certain species, of which the mouse is an example, sinuses are poorly developed. Here the red pulp may be largely cordal.

The cords are bands of tissue separating sinuses. They are made of a reticular meshwork which receives terminating arterial vessels whose meshes are typically filled with free cells conspicuous among which are erythrocytes, macrophages, lymphocytes, monocytes, plasma cells and granulocytes. The adventitial cells of the sinuses send processes out to the contiguous cords and thereby contribute to the reticular meshwork of the cord.

While arterial vessels may in a few instances connect directly with the sinuses, in most instances they empty into the meshwork of the cords. The blood they convey finds its way across the cord, across the wall of the sinus, on through the sinus and its draining veins and out of the spleen. The cords then represent a filtration site not unlike that of the marginal zone. Indeed the marginal zone is continuous with the cords, differing from them in being finer meshed, receiving proportionately more arterial terminations and situated as a shell around the white pulp. The free cells of the cords may be similar in character to those of the marginal zone. Perhaps more erythrocytes are present in the cords.

Blood flow in the spleen is from trabeculae through white pulp and marginal zone into red pulp. This route is the major and dominant vascular pathway, but a backwash exists. Due to venous pressure, a transudate of blood contained in pulp veins is likely to be formed and swept back toward the white pulp in a direction opposite to blood flow.[2] This fluid wave would cross the red pulp, go through the marginal zone, pass on through the white pulp and be collected in the efferent lymphatic trunks girdling the central and trabecular arteries. In the process, it may sweep cells and other materials toward the lymphatics. The movement of small lymphocytes across the wall of sinuses abutting the periarterial lymphatic sheaths and into the white pulp has been noted. The fluid movement toward lymphatics may facilitate the movement of these cells toward the lymphatics. The situation would be analogous to that in lymph nodes. There, small lymphocytes leave postcapillary venules, enter the perifollicular cortex and, if not detained, move out in efferent lymph.[11] In the spleen these cells leave the blood through sinuses which are postcapillary venular vessels, enter that part of the white pulp that corresponds to the extrafollicular cortex of lymph nodes, and exit through lymphatics.

The physiological consequences of this particular microcirculation are manifold. Hazards appear to beset cells in passage in different ways and places in the spleen.

Mechanical Hazards

The passage of cells through the spleen is subject to mechanical hazards at several levels. The endothelium of arterial vessels is noticeably high in the spleen. The endothelium of terminating vessels, as seen in histological sections may be cuboidal and completely efface the lumen. Undoubtedly in life, distended by arterial blood pressure, these arterial lumens would be more capacious. Minimal muscular contraction or swelling of the endothelial cells may, even in life, close off a vessel rather suddenly, or short of that, deform cells passing through the lumen, thereby creating a test of mechanical fragility. Some abnormal cells such as spherocytic or sickled red cells and even normal but aged red cells are mechanically fragile; when they are subjected to mechanical fragility tests in vitro, they are more easily destroyed than young normal red cells. Thus, aged or pathologic cells may not withstand the mechanical test imposed by passage through the spleen.

This type of mechanical hazard is not restricted to arterial vessels. Tortuosities and the irregular narrow passageways of the marginal zone and cords extend the challenge of this mechanical test for passing cells. Under certain circumstances, these pathways may be cylindrical and despite the absence of an endothelial lining, be as efficient and even as stable as actual blood vessels. Vital microscopic observations of Knisely and others would support such a situation. It is also evident that flow through the reticular meshwork of cords and marginal-zone may be neither smooth nor swift, that cells may be held in these places for a long period of time and then released, modified or destroyed. The macrophages and plasma cells set into these meshworks clearly constitute a hazard to cells in passage. It may be noted that many such macrophages and plasma cells were themselves once cells in passage, monocytes or immunologically competent lymphocytes respectively. They were detained in passage through the spleen and underwent transformation into macrophages and plasma cells. As a consequence of their bulk, narrowing the reticular meshwork and making it more tortuous, and as a consequence of their special biological functions (see below), they add to the hazards cells must undergo in passage. The reticular cells of the reticular meshwork may also be phagocytic, the so-called "fixed macrophages," and impose similar hazards upon cells in passage.

Rheological and Metabolic Hazards

Although no blood storage function has been recognized for normal human spleen (in contrast to certain species of animals and in certain pathological conditions in man), it is likely that the hematocrit of splenic blood is considerably higher than that of peripheral blood. Presence of many free cells in splenic pathways may greatly contribute to this elevation of hematocrit. Any preferential skimming of the plasma in the white pulp would elevate the hematocrit still further. Since the viscosity of blood increases as a function of hematocrit, the whole blood viscosity should be higher in spleen than in the general circulation. Further, above the normal hematocrit of 40 to 45 per cent, an increase in viscosity is associated with a decrease in the rate of blood flow. The resulting stasis subjects the cells in passage to splenic hazards for a longer period of time.

Other Physiological Consequences of Stasis of Blood in Spleen

The oxygen-carrying and delivering capacity falls even in the face of a higher hematocrit,[5] giving rise to a relative anoxic state. Some abnormal cells such as those erythrocytes carrying hemoglobin S may succumb to this modest anoxia by sickling, making them more susceptible to other splenic hazards. The glucose level would fall, thereby creating an unfavorable incubation medium especially for red cells which, incapable of oxidative phosphorylation, must depend on anaerobic utilization of glucose. Consequent to the production lactic acid and related metabolites, the concentration of hydrogen ion increases. A fall in pH would cause a shift in the oxygen dissociation curve of hemoglobin, lowering further the oxygen tension.

For red cells, the net result of these physiological changes is a tendency to lose the biconcave shape and become spherocytic.[12] Spherocytes are fragile cells, far more susceptible than normal erythrocytes to the mechanical hazards of passage through the splenic circulation. It should be noted that the stasis of blood in spleen is analogous to incubation in vitro in many of its physiological and biochemical consequences and constitutes an equivalent in vivo of the incubation fragility test in vitro. The term "conditioning" has been used for such an incubation test in vivo which modifies passing cells for further challenge by the splenic circulation.[10]

Immunological Hazards

By its capacity to capture immunologically competent cells of bone marrow and thymic origin, to capture antigen, and to provide a place for their interaction, the spleen is set up to permit the entire series of steps from antigen encounter to antibody production. Should red cells become "not-self" and induce an antibody response, a phenomenon which may occur in phenylhydrazine damaged cells for example, large scale complement-dependent and macrophage-dependent destruction of red cells would occur in the spleen. Even without specific antibody, the presence of immune proteins may induce cellular adherence and facilitate phagocytosis. Thus, incubation of normal red cells of compatible blood groups with the splenic pulp of patients with immune hemolytic anemias causes the red cells to become reactive with Coombs' serum.[18] A formation present in the spleen and marrow, and conspicuous in hemolysis, is a rosette consisting of a central macrophage tightly enclosed by adherent erythrocytes.[1,3]

Macrophages beset cells in passage through the spleen. Indeed, they obtrude upon the passageways and may even block them. Clearly the major determining factor in the destruction of red cells is the condition of the red cell. The spleen may, however, under certain conditions of hypertrophy, as after methyl cellulose administration or chronic passive congestion, take so active a role in destroying cells that normal red cells are destroyed. Here it is likely that the vascular pathways reach so extended and tortuous a condition, and the macrophages accumulate to so high a level that even normal or near-normal cells succumb. The macrophages provide hazards on at least three levels: First, their bulk as affecting filtration has been noted. Second, they are phagocytic and with an unimpeded strike at passing cells, as the spleen provides, they can clearly engulf those cells which are vulnerable. Third, macrophages may secrete hydrolytic enzymes into the pulp. Macrophages contain lysosomes which are rich in acid-hydrolytic enzymes. These catalysts are not only reserved in the cytoplasm to be infixed into phagocytic vacuoles thereby initiating digestion, they are released to the surroundings and thereby provide an enzyme-rich bath for whatever comes into their neighborhood.[17] Undoubtedly dying and dead macrophages contribute to this pool. By histochemistry of the red pulp, one can recognize that certain of those hydrolytic enzymes are not cell-bound but present diffusely through the red pulp with highest concentrations in cords and marginal zone.

Most erythrocytes circulating through the spleen are discharged into cords or marginal zones and then pass into sinuses. The passage from cord into the sinus lumen across the wall of the sinus is a significant movement. Erythrocytes pass alongside the adventitial cells, through fenestrae in the basement membrane and then squeeze between endothelial cells of the sinuses. Should an erythrocyte contain an inclusion, as may be the case after phenylhydrazine or with certain hemoglobinopathies, this inclusion together with some of the cell membrane is pitted out of the cell.[4,9,15] This small portion of the cell is often phagocytized by the sinus adventitial cell or comes to lie in the substance of the basement membrane of the sinus or in the matrix of reticular fibers. The larger part of the cell goes on into the lumen of the sinus and, presumably, on out to the general circulation. A mature cell suffering this membrane loss, unable to synthesize new membrane, loses its biconcave shape and approaches the shape of a spherocyte. Being less deformable than a normal cell, the spherocyte is more easily broken by mechanical stress and consequently remains viable for a shorter than normal span. Under other circumstances, large poorly hemoglobinized erythrocytes may be drawn out through adjacent endothelial slits with the result that portions of some cells are broken off and other cells are hung up over endothelial cells.[4,9,15,22]

THE BONE MARROW

The bone marrow is a complex hematopoietic and lymphatic organ which interacts with the spleen. The primary relationship is that the marrow produces blood cells and the spleen stores them, modifies them or destroys them. Recent work using karyotype or other cell markers has established a broad capacity for cell production by the marrow. The marrow contains a larger number of multipotential cells than other hematopoietic organs; it is the only organ whose cells may mitigate the effects of lethal irradiation. The marrow probably supplies most of the macrophages of the body. The thymus depends upon cells of marrow origin to maintain its population of lymphocytes. The marrow is, perhaps, the source of plasma cells.[13]

The hematopoietic marrow is served by an extensive vasculature. The organizations of the vasculature have been reviewed in detail elsewhere,[21] and only certain aspects need be discussed here. The arterial supply of long bones runs in the central longitudinal axis of the marrow. Radial branches run from the main stem and reach the periphery of the marrow. These branches may be cuffed with lymphocytes, erythroblasts or macrophages in a manner not unlike the periarterial lymphatic sheath surrounding the central artery of the spleen. Venous vessels, the venous sinuses, originate near the periphery of the marrow in or near the encasing bone and run to the central longitudinal axis of the marrow as a complex, communicating spokelike system. In the central longitudinal axis the vascular sinuses run into the major draining venous vessels which run along with the major arterial supply.

It is through the walls of the sinuses that hematopoietic cells exchange between the marrow and the circulation. These walls are thin. They may consist of endothelium alone or of endothelium, adventitial cells and an intervening basement membrane. The layering is dependent upon species and marrow activity. Thus, the rat and rabbit may display the full trilaminar wall, while only an endothelium has been reported in avian marrow. In segments of sinuses in a rat marrow active in the delivery of cells to the blood, the adventitial cells and basement membrane are almost regularly absent and the endothelium attenuated and perforated. Contrariwise, in fatty or gelatinous marrow the adventitial cells are present and voluminous; indeed, they impart the gelatinous or fatty character to the marrow.

The marrow is hematoclastic as well as hematopoietic. Under normal circumstances of hematopoiesis a significant percentage of newly produced cells are defective and are phagocytized. With accelerated hematopoiesis, the level of ineffective hematopoiesis may be much higher, and macrophages, always evident in the marrow, abound. It will be remembered that the bone marrow is the source of macrophages and, therefore, the rapid appearance of great numbers of macrophages in the marow is not unexpected. The marrow contains a large ready reserve of granulocytes and a large ready reserve of reticulocytes. These cells can be released into the circulation with very little delay. Erythroblasts, as they develop into reticulocytes, do so just outside the sinus, pressed upon the abluminal surface of the sinus wall. Thus, the distance traveled by a new reticulocyte into the circulation is a very short one. Granulocytes develop further away and it is reasonable to assume that because mature or nearly mature granulocytes are motile cells, they can make their way over longer distances. Many megakaryocytes lie right against the sinus wall and deliver platelets into the circulation through apertures in the wall. It is known, by the injection of particulate material (for example, carbon and thorium dioxide), that the hematopoietic spaces between vascular sinuses are readily accessible to certain elements of the blood.

REFERENCES

1. Archer, G. T.: Phagocytosis by human monocytes of red cells coated with Rh antibodies. Vox Sang. 10:590, 1965.

2. Barcroft, J., and Florcy, H. W.: Some factors involved in the concentration of blood by the spleen. J. Physiol. (London) 66:231, 1928.

3. Bessis, M.: L'ilot érythroblastique, unité fonctionnellé de la moelle osseuse. Rev. Hemat. 13:8, 1958.

4. Crosby, W. H.: Normal function of the spleen relative to red blood cells. A review. Blood 14:399, 1959.

5. Crowell, J. W., Ford, R. G., and Lewis, V. M.: Oxygen transport in hemorrhagic shock as a function of the hematocrit ratio. Amer. J. Physiol. 196:1033, 1959.

6. Emerson, C. P., Shen, S. C., Ham, T. H.,

Fleming, E. M., and Castle, W. B.: Studies on destruction of red blood cells. Arch. Intern. Med. (Chicago) 97:1, 1956.

7. Goldschneider, I., and McGregor, D. D.: Migration of lymphocytes and thymocytes in the rat. I. The route of migration from blood to spleen and lymph nodes. J. Exp. Med. 127:155, 1968.

8. Harris, J. W.: The Red Cell. Cambridge, Mass., Harvard University Press, 1963, p. 130.

9. Koyama, S., Aoki, S., and Deguchi, K.: Electron microscopic observations of the splenic red pulp with special reference to the pitting function. Mie Med. J. 14:143, 1964.

10. Langley, G. R., and Felderhof, C. H.: Atypical autohemolysis in hereditary spherocytosis as a reflection of two cell populations: Relationship of cell lipids to conditioning by the spleen. Blood 32:569, 1968.

11. Marchesi, V. T., and Gowans, J. L.: The migration of lymphocytes through the endothelium of venules in lymph nodes: an electron microscope study. Proc. Roy. Soc. (Biol.) 159:283, 1964.

12. Murphy, J. R.: Erythrocyte metabolism. III. The relationship of energy metabolism and serum factors to the osmotic fragility following incubation. J. Lab. Clin. Med. 60:32, 1962.

13. Nossal, G. J. V., Cunningham, A., Mitchell, G. F., and Miller, J. F. A. P.: Cell to cell interaction in the immune response. III. Chromosomal marker analyses of single antibody-forming cells in reconstituted, irradiated, or thymectomized mice. J. Exp. Med. 128:839, 1968.

14. Parrott, D. M. V., DeSousa, M. A. B., and East, J.: Thymus-dependent areas in the lymphoid organs of neonatally thymectomized mice. J. Exp. Med. 123:191, 1966.

15. Rifkind, R. A.: Destruction of injured red cells in vivo. Amer. J. Med. 41:711, 1966.

16. Sordat, B., Sordat, M., and Cottier, H.: Localization intra- and intercellulaire d'anticorps spécifiques anti-peroxydase dan les centres germinatifs du ganglion lymphatique poplité de la souris. C.R. Acad. Sci. (D) (Paris) 268:1556, 1969.

17. Sutton, J. S., and Weiss, L.: Transformation of monocytes in tissue culture into macrophages, epithelioid cells and multinucleated giant cells. An electron microscopic study. J. Cell Biol. 28:303, 1966.

18. Wagley, P. F., Shen, S. C., Gardner, F. H., and Castle, W. B.: Studies on destruction of red blood cells. J. Lab. Clin. Med. 33:1197, 1948.

19. Weiss, L.: The structure of fine splenic arterial vessels in relation to hemoconcentration and red cell destruction. Amer. J. Anat. 111:131, 1962.

20. —: The role of the spleen in the removal of normally aged red cells. Amer. J. Anat. 111:175, 1962.

21. —: The structure of the bone marrow. Functional interrelationships of vascular and hematopoietic compartments in experimental hemolytic anemia: An electron microscopic study. J. Morph. 117:467, 1965.

22. Wennberg, E., and Weiss, L.: Splenic erythroclasia: An electron microscopic study of hemoglobin H disease. Blood 31:778, 1968.

23. Williams, G. M., and Nossals, G. J. V.: Ontogeny of the immune response. I. The development of the follicular antigen-trapping mechanism. J. Exp. Med. 124:47, 1966.

Erythrocyte Ion Transport Defects and Hemolytic Anemia: "Hydrocytosis" and "Desiccytosis"

By David G. Nathan and Stephen B. Shohet

ABNORMAL ION TRANSPORT in human erythrocytes has been increasingly recognized as a concomitant feature of several forms of hemolytic anemia.[67] The conditions in which disorders of ion transport have been demonstrated are reviewed in this paper in an attempt to define the pathophysiological role of these transport defects in the reduction of red cell survival. It should be stressed that a great deal of conjecture still surrounds opinions concerning the significance of ion transport defects in hemolytic anemia. Despite rapid advances in the field of membrane physiology, awareness of the influence of the ion transport system as it is studied in vitro with the interaction of red cells and the reticuloendothelial system in vivo is still too incomplete to permit more than cautious estimates of pathophysiologic mechanisms. The evaluation of these mechanisms is of practical as well as theoretical importance. Improved management of patients with hemolytic anemia and enhanced red cell storage capability in blood banks depend upon firm understanding of the relationships which exist between alterations of red cell membrane chemistry and survival in vivo.

Accumulation of sodium in excess of loss of potassium from human cells in vitro is associated with an obligatory gain of cell water and, hence, a shift toward the critical hemolytic volume ("hydrocytosis").[123] This alteration in volume is thought to provoke an increased rate of destruction in vivo. Maintenance of normal cellular hydration, however, is not a requirement for normal survival in vivo after storage.[96,97] Recent studies of patients with congenital abnormalities of red cell ion transport[113,162] have shown that large increments of steady state ion concentrations may be compatible with virtually normal erythrocyte survival. It seems likely, then, that some ion abnormalities which occur during red cell storage are actually only epiphenomena which accompany much more important, if more subtle, deviations of metabolic and membrane integrity. On the other hand, the ion transport abnormalities which lead to cellular dehydration ("desiccytosis"), such as those observed in sickle cell anemia[50] and pyruvate kinase deficiency,[87] may play a central role in the pathophysiology of these diseases.

From the Division of Hematology of the Department of Medicine, Children's Hospital Medical Center and the Department of Pediatrics, Harvard Medical School, Boston, Mass.

Supported by USPHS Grant HD 02777, TIAM 05581, and by a grant from the John A. Hartford Foundation, Inc.

David G. Nathan, M.D.: *Chief, Division of Hematology, Children's Hospital Medical Center; Assistant Professor of Pediatrics, Harvard Medical School, Boston, Mass.; recipient of USPHS Career Department Award K 3AM35361.* Stephen B. Shohet, M.D.: *Associate in Hematology, Children's Hospital Medical Center; Assistant Professor of Pediatrics, Harvard Medical School, Boston, Mass.; Fellow of the Medical Foundation, Inc.*

In order to review the clinical disorders of red cell function which illustrate the role of ion transport in cell survival, it is necessary to consider a brief summary of the pertinent features of normal human red cell ion physiology.

WATER AND ELECTROLYTE METABOLISM IN HUMAN RED CELLS

The human red cell is a biconcave bag of hemoglobin, enzymes, glycolytic intermediates and glycolytic cofactors dissolved in a mixed salt solution. The pH of the solution is controlled by hydrogen ion flow from the plasma, as well as by the production of three-carbon acids during glycolysis. Buffering capacity is provided by the amino acids of the hemoglobin and by the carbonic acid: bicarbonate system.[19] Electrochemical and osmotic equilibria between red cells and plasma are achieved by the ion transport characteristics of the membrane. Univalent anions are quite freely exchanged between the plasma and the cells,[116] obedient to the dictates of the Donnan effect.[31,145] An energy-dependent cation pump[101] counters the slow leakage of univalent cations between these compartments. The oncotic pressure exerted by the red cell hemoglobin and of the relatively impermeable organic phosphates is balanced by the pump-dependent maintenance of a lower cation concentration within the cell water than in the surrounding plasma water.[53,143] Arrest of the activity of the cation pump and resultant equalization of cell and plasma cation concentrations inexorably leads to cell swelling and to eventual hemolysis in vitro because the extra oncotic pressure exerted by the internal hemoglobin and organic phosphates creates a net flux of water into the cell.[19,100,111,137,153] This rather elaborate membrane transport system is required because of the gracile structure of the membrane itself. It is flexible and plastic enough to afford the cell the suppleness which it requires to traverse narrow and tortuous microcirculatory passages. A more rigid and impermeable structure, though less dependent upon complex pumps and permeability characteristics, would constitute a preposterous encumbrance to red cell navigation.

The molecular basis of ion and water transport through red cell membranes has as yet defied precise definition. In fact, the bilayer structure of the membrane proposed by Davson and Danielli[26,28] which serves so admirably as a model for characterization of membrane transport phenomena may not represent actual structural conditions.[56] The membrane is comprised of multiple lipo- and glycoproteins,[115] insofar as acrylamide gel electrophoresis can separate these components.[127] Interwoven with these proteins are structural phospholipids, mainly phosphatidyl choline, and phosphatidyl ethanolamine and sphingomyelin.[144] These phospholipids and proteins are probably arranged in a reasonably orderly fashion, depending in part upon the solubility characteristics of the ends of the lipid molecules. Some phospholipid molecules, as well as cholesterol, are thought to be in the outer layer of the membrane and more loosely attached to the inner core of lipoprotein because they are in relatively rapid equilibrium with plasma lipids.[125,128] Complex carbohydrates, which afford many of the antigenic determinants, also tend to localize relatively near the plasma surface of the membrane.[27,146] Water, small molecules (the passage of some of which is facilitated by special transport systems), and ions permeate through the interstices of these molecules. As mentioned above, anions traverse the membrane readily and water achieves nearly instant equilibrium, but cation permeability is markedly retarded. These facts have led to the proposal of a theoretical "pore" through the membrane,[138,140] the lining and stoma of which must be positively charged. The nature of this positive charge in the membrane is not firmly established. Passow[116] has proposed that amino groups of membrane proteins are the most likely sources of this charge. The charges on amino groups are affected by changes in pH, and pH changes are known to influence ion permeability in a predictable manner. For example, a rise in intramembrane pH would tend to reduce the positive charge on amino groups. Indeed, cation permeability is enhanced and anion permeability reduced by exposure of red cells or ghosts to alkaline buffers. The reverse occurs at an acid pH. Another model of membrane charge barrier is based upon the consideration of the divalent cation content of membranes.[12,42,84,139] Although the magnesium content of erythrocytes is very difficult to eliminate, there is evidence that red cell calcium can be removed almost entirely by exhaustive dialysis in the presence of EDTA.[51] The charge associated with calcium in the membrane normally might impede the progress of univalent cations such as sodium or potassium through the pores. The role of Ca^{++} in membranes, however, is very complex. Under some conditions, it prevents excessive cation loss;[12,84] in others it induces marked membrane leakiness.[42,54,77] Whatever may be the basis of the electrostatic barrier to ion permeability, the result is that univalent cations move relatively slowly across the cell membrane in response to the concentration gradients which drive them. In the event of damage to the pore barrier, potassium loss immediately precedes sodium gain with resultant cell shrinkage because in the normal steady state:

$$\frac{[K]_i}{[K]_o} \text{ is greater than } \frac{[Na]_o}{[Na]_i}$$

If the damage persists, sodium accumulation usually occurs and the cells swell prior to hemolysis. Some types of damage, such as the injuries induced by deoxygenation of sickle cells, by cyanide treatment of pyruvate kinase-deficient cells or by fluoride treatment of normal cells, induce continued excessive potassium loss without equivalent sodium gain. Cell shrinkage persists in these situations.

To deal with the cations which leak across the cell, an energy-dependent, membrane-localized cation pump, the energy of which is derived from membrane ATPase activity,[120] returns them to their original levels. Thus, the pump drives sodium against a concentration gradient from the cell back to the plasma, and potassium from the plasma to the cell. The membrane ATPase requires magnesium and is further activated by sodium and potassium. Approximately 70 per cent of the sodium:potassium activated component is inhibited by ouabain (pump 1) and approximately 60 per cent of the remainder is inhibited by ethacrynic acid (pump 2).[54]

The influence of the cation pump on red cell metabolism has been analyzed in several laboratories.[35,91,92,100,114,131,132,163] The pump accelerates glycolysis and ATP production, probably by providing ADP for the red cell phosphoglycerate kinase reaction for which ADP availability appears to be limiting[35,114,163] and by providing inorganic phosphate for the hexokinase and phosphofructokinase reactions.[91,92] Ouabain inhibits ADP and Pi production from ATP by ATPase; it slightly depresses lactate production in intact erythrocytes and in hemolysates in which membranes are present. The fraction of glycolysis devoted to the ATPase pump is not known and remains a problem central to the understanding of deranged red cells. Whittam[159] has calculated that only approximately 8.5 kilocalories per hour are required for active transport of potassium and sodium, whereas approximately 70 kilocalories per hour are produced by erythrocyte glycolysis if all of the glucose consumed is converted directly to lactate without entering into the Rapoport-Luebering cycle.[122] Indeed, arrest of the pump with ouabain[100] or its stimulation by amphotericin[11] alters lactate production by erythrocytes only minimally. Until more is known about the fraction of newly synthesized 1, 3-diphosphoglycerate (1, 3-DPG) which is handled by 2, 3- DPG mutase or phosphoglycerate kinase, the influence of the pump on the ATP economy in erythrocytes will be poorly understood.[153]

In hemolytic anemias in which reticulocytes and young red cells abound in the peripheral blood, functional abnormalities of ion transport are difficult to define. Human reticulocytes contain mitochondria but they do not utilize oxidative pathways unless glycolysis is blocked either by a drug such as fluoride or by an enzymopathy such as pyruvate kinase deficiency.[72] Glycolysis in reticulocytes is greatly increased and the level of 2, 3-diphosphoglycerate (2, 3-DPG) is elevated when compared to normal cells.[8,45,122] ATP turnover is accelerated. Membrane ATPase activity, cation concentrations and flux rates are also increased in young cells,[8,10,15,69] although the concentrations of cations tend to remain fairly stable during incubations in vitro.[105] Therefore, studies in vitro of the young red cells of patients with hemolytic anemia must be accompanied by appropriate cell–age adjusted controls. Such age matched controls can be difficult to acquire, particularly for comparison with cells having heterogeneous survival rates. In fact, measurements of ion flux in the cells of patients with hemolytic anemia almost always reveal multiple rates of ion transport which probably result as much from the heterogeneity of the cells[154] as from different compartments within individual cells.[137]

INFLUENCE OF CELL ENVIRONMENT ON CELL PERMEABILITY AND CATION PUMPING

The Plasma

Whittam has recently reviewed environmental influences on ion transport in red blood cells.[155] Certain aspects of this subject are of specific importance in our understanding of clinical disorders of red cell permeability. Sodium and potassium fluxes occur at rates of about 3 and 2 mEq./L. cells/hour, respectively. Electrochemical neutrality is maintained by transport of a chloride ion with the extra sodium. The passive leaks of these ions are quite independent but the active transport processes are linked. Thus, active sodium efflux is arrested if potassium or a potassium substitute such as rubidium is absent from the medium. Both active fluxes are enhanced by an increase in pH or in intracellular sodium concentration or by a decrease in intracellular potassium concentration.

Plasma albumin does not appear to influence ion flux very significantly. Even though red cells incubated in pure salt solutions develop surface spicules which are abolished by relatively small concentrations (0.2 mM) of albumin, neither potassium nor sodium flux is clearly influenced by this shape change alone. Complement fixing antibodies create major flux alterations in red cells.[43,44] Sodium gain exceeds potassium loss in complement injured cells with the result that swelling and hemolysis ensue in vitro. We have not observed any metabolic or transport abnormalities in red cells coated with incomplete warm antibody without complement fixation, although Schrier and his co-workers have shown that red cell ghosts coated with an anti-D antibody synthesize ATP less readily than do uncoated ghosts.[133]

Calcium and Magnesium

The effects of unbound calcium and magnesium in the medium of suspended red cells have been the subject of recent inquiries by Weed, LaCelle and Merrill.[149] Membrane stiffness appears to increase if calcium, and to a lesser extent magnesium, accumulate above normal levels within the cell membrane. Intracellular ATP prevents the stiffness associated with excess intramembrane calcium, presumably by chelation of the calcium. Lack of this chelation effect of ATP may contribute to the increased cell rigidity observed in certain hemolytic anemias or in aged normal red cells in which ATP levels are depleted. EDTA prevents calcium accumulation in membranes and also reduces membrane stiffness in intact cells or ghosts. In addition to the development of membrane stiffness, ATP depletion and consequent calcium ingress into red cell membranes are associated with markedly enhanced potassium permeability, water loss and cellular dehydration. These effects of ATP depletion are inhibited by EDTA or to a lesser extent by removal of calcium from the medium. The destruction of pyruvate kinase deficient erythrocytes may be intimately related to this mechanism of red cell membrane injury (see below).[87]

Drugs and Metabolic Inhibitors

Inhibitors of red cell metabolism or membrane function lead to alterations in cation permeability which may be instructive as models for certain clinical disorders.

Fluoride inhibits enolase and, therefore, decreases red cell ATP. At high concentrations (10 mM) and in the presence of calcium ion, the drug also damages membranes and produces EDTA-inhibitable potassium loss and cell shrinkage.[37,42,77] The cation pump ceases not only because of loss of ATP but also because fluoride completely abolishes Na^+, K^+ and Mg^{++} ATPase activity even in the presence of excess magnesium.[37] Iodoacetate also inhibits red cell glycolysis and, at high concentration, in addition provokes cell shrinkage and potassium loss.[156] The basis of the membrane effect of iodoacetate has not been clearly established.

Blockade of membrane thiols leads to increased ion leak,[57] probably secondary to disturbances of the normal apposition of membrane proteins. Polyene antibiotics such as amphotericin or fillipin disrupt the bonds holding cholesterol to the membrane proteins.[73,74] Potassium and sodium leaks are markedly increased beyond the capacity of the cation pump to oppose them. Red cells exposed to such drugs, therefore, swell rapidly. The effects of amphotericin are reversible, however. For this reason, patients treated with intermittent intravenous doses of this drug do not develop hemolytic anemia.[11]

Hydrogen peroxide attacks the internal thiols of the red cell, including glutathione and globin itself. H_2O_2 is disposed of by red cell catalase[59] and glutathione peroxidase.[21] The oxidized glutathione which is produced non-enzymatically or via glutathione peroxidase[21,89] is reconstituted by glutathione reductase. Hence, hydrogen peroxide production is intimately related to the hexose monophosphate shunt via the TPNH requirement exerted by glutathione reductase.[20,61] Red cell membrane protein thiols are probably also attacked by hydrogen peroxide when glutathione is depleted, as in glucose-6-phosphate dehydrogenase (G-6-PD) deficiency. However, membrane resistance to peroxidation is high, due to the presence of fat soluble vitamin E which serves as a sump for free radicals and thus protects vital membrane structures from peroxidation.[32] Vitamin E-deficient red cells exposed to hydrogen peroxide undergo rapid lysis which is preceded by potassium loss and cell shrinkage, followed by sodium plus water accumulation and cell swelling.[82] Peroxidation of unsaturated fatty acids of membrane phospholipids (predominantly in phosphatidyl ethanolamine) may occur prior to peroxide induced hemolysis in vitamin E deficiency.[63,82] Actual loss of phosphatidyl ethanolamine develops only when the vitamin E deficiency is very severe and the peroxidative insult very large or prolonged.[63] With moderate peroxidative threats, saturated (peroxide resistant) fatty acid derived from neutral lipids and phosphatidyl choline may replace the oxidized fatty acid in the membrane phosphatidyl ethanolamine.[82] Whether the phosphatidyl ethanolamine is lost or not, H_2O_2 does hemolyze vitamin E-deficient red cells. Presumably either the peroxidative attacks on membrane proteins and/or the change in fatty acid composition of phosphatidyl ethanolamine are responsible for both the cation leak and the hemolysis.

Oxidant drugs such as primaquine[22,147] or methylene blue also affect red cell membrane permeability and enhance potassium loss and pumping. Before such positively charged agents can enter the cell they must be reduced by a membrane diaphorase which utilizes either DPNH or TPNH as a hydrogen donor.[129,130] If the reduced forms of these drugs are subject to autoxidation after their entry, hydrogen peroxide may be formed and an attack on intracellular and membrane thiols can ensue.[22] When this damage occurs, increased cation flux may be accompanied by Heinz body formation.[30] Direct oxidation of membrane and hemoglobin thiols may also proceed in the absence of drug autoxidation.[57]

The enzymes contained in certain snake venoms or bacterial exotoxins may be responsible for severe cation leak and osmotic lysis of red cells. Phospholipase A_2 frees fatty acid from the beta position of phospholipids and creates lysophosphatides which form micelles in preference to bimolecular leaflets. The micelle formation is probably associated with increased intramembrane water which in turn may permit increased cation permeability, cell swelling and osmotic lysis.[4,29]

Ouabain and the cardiac glycosides are antagonists of sodium, potassium activated ATPase activity. The difference between total potassium influx or sodium efflux and the same fluxes during exposure to ouabain constitutes the active cation pump. Ouabain reduces glycolysis slightly[101] in mature red cells (see above) and may markedly reduce glycolysis in cells with very active cation pumping (see below).[162] Exposure to ouabain leads to gradual cell swelling and eventual hemolysis.

Effects of Circulation

As described above, red cell metabolism and cation balance, and, therefore, the state of cell hydration, are markedly affected by alterations of the pH and the ionic composition of the medium, as well as by the presence of certain drugs. Intracellular alterations also affect these functions. Glycolysis[1] and cation fluxes[109] are both increased by the deoxygenation of red cells. These effects of deoxygenation are not entirely understood. They may be related to the binding of organic phosphates, including 2, 3-DPG and ATP by the deoxygenated hemoglobin.[5,6,16] The binding of ATP may release its inhibition of phosphofructokinase, which is an important control point for red cell glycolysis.[81] Internal pH shifts accompany deoxygenation because hydrogen ions are then bound by the hemoglobin [the alkaline Bohr effect[117]]. The resultant rise in intracellular pH would enhance both glycolysis[90,93] and cation fluxes.[157]

The splenic environment may be characterized as hypoxic,[68] static[36,47,152] and acidotic.[68,102] The metabolic effects of hypoxia described above are countered by external acidosis which diminishes glycolysis[90,93] and cation flux.[157] Stasis may create even more membrane injury. The membranes may be ingested by splenic littoral cells, if antibody is present on the red cells;[79] decreased membrane lipid renewal may occur when the stasis is complicated by acidosis;[99] and finally splenic phospholipase may directly attack red cell membrane lipids.[13,41,78] Reticulocytes with metabolic blocks along the glycolytic pathway, such as in pyruvate kinase deficiency, are particularly threatened by long sojourns in the hypoxic spleen,[7,65] since the pO_2 of splenic blood is well below that required for reticulocyte mitochondrial function. Such cells rapidly lose ATP, potassium and water and develop severe hyperviscosity.[87] The details of this lesion will be described further below.

CLINICAL DISORDERS OF CELL MEMBRANE PERMEABILITY

Cells with Smooth Surfaces

Hereditary spherocytosis. Hereditary spherocytosis (HS) was the first of the red cell defects in which an ion transport abnormality was detected. The basic abnormality of the HS cell, however, remains unknown. HS is the most common of the congenital hemolytic anemias. Several new cases are observed annually in a large pediatric center and the number of mild cases which escape detection is probably much greater. The disease appears to be due to an abnormality of red cell membrane formation and stability, although a subtle error in red cell energy metabolism leading to a secondary membrane defect has not been ruled out entirely. The membrane defect is accompanied by instability of membrane lipids, a subject which has been discussed previously by Cooper.[25a]

Harris and Prankerd first described abnormal sodium permeability in HS.[48] Bertles further explored the characteristics of sodium permeability in HS cells[9] and found that in approximately half of his cases there was an elevation of sodium influx to as high as 6.1 mEq./L. cells/hour, the normal range being 2.8–3.8 mEq./L. cells/hour. Jacob and Jandl extended these observations in their comprehensive study of HS cell metabolism and permeability.[58] They confirmed that HS cells were slightly more permeable to sodium than were normal cells. In two experiments, HS cell sodium influx was 4.6 and 4.8 mEq./L. cells/hour compared to control values of 2.3 and 2.6 mEq./L. cells/hour. In other experiments, they noted that sodium efflux was also increased in HS cells to 5 mEq./L. cells/hour compared to 3.7 mEq./L. cells/hour in normal cells. More recently, Wiley confirmed these observations on sodium pumping and also showed that ATPase activity was increased in postsplenectomy HS cells.[161] The ATPase activity was, in fact, linearly related to the rate of sodium efflux in HS cells,[161] but neither the ATPase activity nor the sodium efflux was related to their life span.[160]

Jacob and Jandl have also observed that potassium flux is normal in HS cells. This observation implies that HS cells should gain potassium while maintaining a normal sodium concentration during incubation in vitro. If passive sodium influx is excessive and passive efflux of potassium is normal, increased active sodium efflux should maintain the internal sodium concentration constant, while the increased linked active potassium influx should cause potassium accumulation. Indeed, incubation of HS cells, particularly presplenectomy but also postsplenectomy, reveals a progressive accumulation of potassium in the cells without a rise in cell sodium.[109]

Jacob and Jandl further explored the metabolism of postsplenectomy HS cells by evaluating their glucose consumption at high and low phosphate concentrations without and with ouabain, 3.5×10^{-5} M. The HS cells had a higher basal rate of glucose consumption than did normal cells and this rate was significantly diminished ($-18.5\% \pm 4.0$) by ouabain. They observed no significant change in glycolysis in control cells incubated with ouabain. The authors concluded that a larger than normal fraction of cell energy was devoted to ion pumping in HS cells and they further showed that deprivation of glucose from HS cells resulted in excessive sodium accumulation, cell swelling and rapid splenic sequestration.[58] Mohler[95] added the observation that HS cells lost ATP more rapidly than normal cells when they were deprived of glucose and that ouabain slowed the rate of ATP loss. All of these data influenced the authors to conclude that a major and perhaps primary defect in HS cells might be excessive sodium permeability. They proposed that the resultant demand for increased glycolysis and pumping might not be met in the unfavorable environment of the spleen and that the cells would eventually lyse there, if prolonged sequestration induced glucose deprivation.

On the other hand, the actual change in sodium permeability in HS cells is very small and the increment of glycolysis required to meet its demand is also small. HS cells also tend to be dehydrated, rather than watery.[83] Their cation and water contents are lower than normal and they are actually more viscous than normal cells, particularly at a pH range between 6 and 7[66,102] [the range thought to be present in the spleen[68]]. The increased glycolysis in HS cells may be due in part to a decrease in their mean age, as well as to their slightly increased rate of ion pumping. Splenectomy almost always cures the anemia and reticulocytosis observed in HS but diisopropylfluorophosphate (DFP) ^{3}H survival studies show that the life span of HS cells, postsplenectomy, is 96 ± 13 days, rather than the normal 123 ± 14 days.[18] For this reason, the rate limiting and age dependent glycolytic enzymes (hexokinase, phosphofructokinase, pyruvate kinase), as well as membrane ATPase, are all apt to be somewhat increased in activity in postsplenectomy HS blood.[8,15,17,45,94,122] In addition, age related glycolytic intermediates and cofactors, which are known to influence the overall rate of glycolysis might be expected to be present in variable proportions in postsplenectomy HS cells of variable age. The relative youth of the HS cells might also contribute somewhat to the observed permeability changes since ion flux is considerably greater in a young red cell population.[8,69] It is unlikely that the spheroidal shape itself is responsible for increased glycolysis. Indeed, Cooper and Jandl created spheroidal cells from normal cells by removing 37 per cent of the surface cholesterol and detected no change in glycolysis.[24] Further, the authors' failure to detect any change in glycolysis in normal cells exposed to ouabain, though confirmed by some,[158] has not been the experience in all laboratories where, in fact, the phenomenon has been regularly observed in normal cells.[101,109] Finally, patients with high sodium red cells and strikingly increased permeability and pumping have recently been discovered.[113,162] Some do not have hemolytic anemia at all. The influence of small increments of pumping on the life span of red cells, therefore, seems to be minor.

Jacob extended the observations on HS cell pumping in another interesting direction. Stimulated by the observations of Hokin and Hokin[55] on the possible role of phosphatidic acid in sodium pumping, he examined the hypothesis that the accelerated pumping observed in HS cells might actually cause their increased membrane lipid loss.[60,62] He noted that the cells lost phospholipid during incubation and, most importantly, that the loss was inhibited by ouabain. He also noted an increase in the ^{32}P labeling in HS cells which was also inhibited by ouabain. Cooper and Jandl,[23,24] however, showed that HS cells lost phospholipid only in their agonal stages, if they were maintained in an environment which contained serum as a source of lipoproteins. Shohet has noted that the turnover of red cell phospholipid as measured by acylation of radioactive fatty acid was consistent with cell age in HS.[136,162] In addition, the recently discovered patients with high red cell sodium had no detectable alterations in red cell lipid stability or turnover and their cells were not spheroidal.[131,162]

Thus, the HS cell continues to elude precise understanding. It is spheroidal and it is trapped mainly in the spleen. Most workers are in agreement with these facts.[49,121,148] Cooper has reviewed the data[24] which have resulted from various studies of lipid metabolism in HS cells. One of the most interesting of these is the fact that although the "fragments"[124,148] which fall off the HS cell during prolonged incubation do not contain protein,[76] the established reduction in HS cell surface area cannot be explained by lipid deficiency alone. Thus, the geographic disposition of the lipid in, or the protein conformation of, the HS cell membrane may be responsible for the reduction in surface area. In any case, the HS membrane poorly maintains its slippery grasp on cholesterol and even loses its phospholipid if it becomes severely deprived of an energy source. The observed abnormalities of glycolysis and cation transport are more apt to be secondary rather than primary manifestations of this membrane disturbance.

Elliptocytosis. Peters and his associates[118] performed studies of sodium efflux from the red cell ghosts of 13 patients with hereditary elliptocytosis. In nine of these patients, the reticulocyte count was higher than 2 per cent and sodium efflux was definitely increased above normal. The authors concluded that their data suggested a membrane lesion in the elliptocytic membrane, but the elliptocyte cell population was young relative to the normal control cells and, therefore, membrane ATPase activity was probably increased.[15] It has, therefore, been difficult to be certain whether the observed increase in sodium efflux constituted a primary or a secondary manifestation of the disease.

Stomatocytosis. Occasional red cells with a slit-like, instead of a circular, area of central pallor may be observed in normal blood smears. On some occasions, however, this alteration has been observed in a large proportion of the cells and hemolytic anemia has been present. Lock and his co-workers[80] observed the first reported case of hemolytic anemia with stomatocytosis and this report was followed several years later by the case studies of Meadow[85] and Miller and his co-workers.[88] In none of these patients were red cell electrolytes measured. Recently, Zarkowsky and his co-workers[162] reported a 3½ year old boy of Hungarian descent with splenomegaly, congenital hemolytic anemia (hemoglobin 6 Gm. per 100 ml.) and jaundice (bilirubin 5 mg. per 100 ml.). Prior to splenectomy, his red cell indices were MCV 151 cu. μ, MCH 35 $\mu\mu$g., and MCHC 24 Gm. per 100 ml. Reticulocyte count ranged from 20 to 40 per cent and at approximately 1 year of age a splenectomy was performed. His hemoglobin has remained at approximately 10 Gm. per 100 ml. with 10 per cent reticulocytes ever since. The MCV was now 118 cu. μ, MCH 27$\mu\mu$g. and MCHC 23 Gm. per 100 ml. The peripheral smear showed that the red cells were large and that many had slit-shaped areas of central pallor. Wet preparations showed that the red cells were bowl-shaped instead of biconcave. No spherocytes were seen. Osmotic fragility was increased as was autohemolysis. The increased autohemolysis was not reduced by ATP or glucose. Maximal swelling of the red cells occurred in 0.6 per cent saline when their volume was 182 cu. μ, compared to a normal maximal value of 158 cu. μ. Thus, the critical hemolytic volume[46] of these cells was increased, indicating a large membrane area, but the initial volume of the cells was also increased indicating a high red cell water content. In fact, the water content of the cells was greatly increased being 77.7 per cent of the red cell volume, a normal control being 68 per cent. A second study of cell water performed after 5 hours of incubation in autologous plasma at ambient temperature was 72.9 per cent. The high water content of these cells suggested an abnormality of membrane permeability. Table 1 presents the extraordinary results of the metabolic and cation flux studies which were obtained. The cells had an enormous sodium content and a markedly decreased potassium content as well. The passive leaks of these ions were vast and were incompletely countered by active transport rates of approximately 10 times normal. The rates were of such magnitude that the precise values must be considered estimates because of technical limitations. The fraction of glycolysis devoted to active transport was also extremely high. As much as 2.7 mM of glucose/L. cells/hour were utilized for transport. This figure amounts to the turnover of 5.4 mM of ATP/L. cells/hour for active transport, compared to 0.5 mM of ATP/L. cells/hour utilized for this purpose in normal cells. This 10-fold increase in ATP turnover was carried out by a total membrane ATPase activity which was no higher at maximum velocity than that of cells of similar age with ordinary rates of pumping[164]. Since the pumping was performed against much lower concentration gradients than were present in normal cells, the actual work of pumping per molecule transported was lower in the patient's cells than in normal cells. Since the ratio

$$\frac{\text{ATP utilized per liter of cells per hour}}{\text{ions transported per liter of cells per hour}}$$

was similar in patient and control cells, the efficiency of pumping was considerably lower in the patient's cells.

During the studies of the influence of ouabain on glycolysis and pumping in the patient's cells, a peculiar additional abnormality was noted. ATP declined rapidly in these cells when ouabain was present. This finding was unexpected since inhibition of ATPase would be expected to preserve intracellular ATP. Clearly, the inhibition of ATPase also reduced the formation of a product essential for the maintenance of intracellular ATP. Further investigations of intracellular glycolytic intermediates and cofactors and their influence on the kinetics of glycolytic enzymes must be performed in this

Table 1.—Sodium and Potassium Fluxes and Intracellular Cation Concentration

	Glucose Consumption (mM/L. Cells/Hr.)	ATP Zero Hours (μMole/L. Cells)	ATP 3 Hours (μMole/L. Cells)	Sodium (mEq./L. Cells)	Potassium (mEq./L. Cells)	Potassium Flux (mEq./L. Cells/hour) In	Out	Net	Sodium Flux (mEq./L. Cells/hour) In	Out	Net
Patient	4.8	1100	1100	91–108	35–47	31.0	30.0	+1.0	107.0	89.0	+18.0
+ Ouabain	2.1	1100	300			16.0	32.0	−16.0	100.0	68.0	+32.0
ATPase pump	2.7	—	—			15.0				21.0	0
Normal Control*	1.6	1500	1800	10±6	100±4	1.7	1.7	0	2.7	2.7	0
+ Ouabain	1.35	1500	1500			0.4	1.5	−1.1	2.7	1.0	+1.7
ATPase pump	0.25	—	—			1.3				1.7	+1.7
Patients with hereditary spherocytosis†						2.5	1.8	+0.7	4.7	4.7	0
Patients with pyruvate kinase deficiency†						3.6	4.7	−1.1			

*Normal values for sodium and potassium flux representative of many determinations in our laboratory and other published data; standard deviation for these values, ±10 per cent.
†Mean values of four such patients.

patient's cells in an effort to determine the nature of this defect. As mentioned earlier, the cation pump influences glycolysis in red cells because the ADP produced by membrane ATPase stimulates the phosphoglycerate kinase (PGK) reaction in the cytosol and the inorganic phosphate produced stimulates hexokinase and phosphofructokinase. An increased requirement for ADP by the PGK in this patient's cells might have led to ouabain-induced reduction of PGK activity because of reduced production of ADP by the membrane in the presence of this inhibitor.

The basis of the defect in cation permeability in this patient was not established, despite a number of other studies of membrane function. Investigations of "pore charge" or "pore size" utilizing penetration rates of divalent anions,[116] divalent cations, a univalent cation (Cs^+), 4 and 5 carbon alcohols[140] and water[134] showed that the cells behaved normally with respect to penetration of the substances, except for the univalent cation. The reflection coefficients[71] for thiourea, urea and malonamide were also normal. These results supported the view that the theoretical pores were of normal size and number. Alterations in the pH of the incubation media shifted both sodium and sulfate permeability in the expected opposite directions. These observations implied that a normal pore charge was present in the membranes and that the massive cation permeability was due to an entirely different mechanism as yet unknown.

Finally, studies of lipid metabolism in these cells failed to show any significant abnormality. Fatty acid incorporation into membrane phospholipids was normal for cell age[38,136] and there was no detectable phospholipid loss during a 3-hour incubation. As mentioned above, these results suggested that there was no direct association between the cation pump rate and phospholipid turnover in human red cells.

Shortly after this patient was reported, other patients with stomatocytosis and hemolytic anemia were described. In one case reported previously to us by Dr. Daniel Lane and Dr. William B. Castle,[75] mild hemolytic anemia and stomatocytosis were present without alteration of cell electrolytes. More recently, Norman,[110] Muir-Jackson and Knight[98] and Ducrou and Kimber[33] have emphasized that stomatocytosis and hemolytic anemia were common in individuals of Mediterranean descent in Australia and that red cell electrolyte composition was normal in these cases. Clearly, stomatocytosis is a nonspecific finding, independent of cell electrolyte composition; yet its presence may signify a cell electrolyte and permeability abnormality. It was, in fact, on the basis of the observation of stomatocytosis that Naiman discovered a second case of high sodium red cells in a 4-year-old boy who was studied only because the diagnosis of HS had been made in his father. The boy was entirely well. although he was slightly anemic. The propositus, his father and his paternal grandfather were studied and reported by Oski and his coworkers.[113] Interesting comparisons were made with the red cells of the first patient with high sodium red cells reported by Zarkowsky and his coworkers.[162]

Most of the studies were performed on the blood of the 37-year-old father of the propositus (E.S.) who reported transient episodes of jaundice since the

Table 2.—Sodium and Potassium Fluxes and Intracellular Cation Concentrations

Subject	Potassium Flux (mEq./L. Cells/Hour)			Sodium Flux (mEq./L. Cells/Hour)			Cation Concentration (mEq./L. Cells)	
	In	Out	Net	In	Out	Net	Sodium	Potassium
P.S., propositus	3.2	3.8	−0.6				59.3±1.2	73.5±2
+ ouabain	0.6	3.7	−3.1		Not Done			
ATPase pump	2.6							
E.S., father	3.3	3.6	−0.3	16.8	14.7	+2.1	58.5±2.7	8.5±1.2
+ ouabain	0.5	3.4	−2.9	15.9	7.8	+8.1		
ATPase pump	2.8				6.9			
J.S., grandfather	3.5	5.5	−2.0	13.9	15.3	−1.4	50±1.8	86±1.7
+ ouabain	0.7	5.0	−4.3	14.3	7.3	−7.0		
ATPase pump	2.8				8.0			
Normal	1.7	1.7	0.0	2.7	2.7	0.0	10±4	100±4
+ ouabain	0.4	1.5	1.1	2.7	1.0	+1.7		
ATPase pump	1.3				1.7			

age of 19. Splenectomy was performed at age 32 because of jaundice and increased red cell osmotic fragility. Stomatocytosis was present in the propositus and his father, but not in the grandfather. Osmotic fragility was increased in all three. The MCV was also increased and autologous red cell survival was slightly short as measured with ^{51}Cr. There was only mild anemia and no reticulocytosis. Table 2 shows the results of sodium and potassium flux studies. There was a moderate reduction in cell potassium and a marked increase in cell sodium. Total cation content was increased. Active transport of potassium was increased twofold, whereas active sodium transport was increased three to fourfold. Sodium leak was increased sixfold. These results suggested that the major red cell membrane lesion was an increased sodium permeability with increased pumping that was insufficient to reduce internal sodium to normal values. Centrifugation of the red cells to provide age-dependent separations, however, revealed important additional and unanticipated information. Normally, young red cells, which are less dense than old red cells, accumulate in the top of a column of centrifuged blood, whereas old cells are found in the bottom. The cells of E.S. behaved in a paradoxical fashion. Table 3 shows that the top fraction contained old cells with no reticulocytes and relatively low transaminase (GOT) activity, whereas the bottom layer contained young cells with relatively high GOT activity and reticulocytes. The old cells were less dense than the young cells because they accumulated sodium. They also lost potassium, although their total cation content was greater. Therefore, they gained water, which gave them a greater buoyancy. As expected (Fig. 1), osmotic fragility was also greater in the top (old) cells than in the bottom (young) cells. Investigations of pump activity revealed that although the young red cells had an increased sodium content and leak rate, the increased leak was almost completely compensated for by

Table 3.—A Comparison of the Properties of Young (bottom 10 Per Cent fraction) and Old (top 10 Per Cent fraction) Red Cells from E.S., the Father.

	Top Fraction	Bottom Fraction
Reticulocytes (%)	0.0, 0.0	6.5, 4.1
MCV (cμ)	116.4	92.0
MCH (μ μg./cell)	30.6	28.4
Stomatocytes (%)	32.0	0.0
Red cell GOT (U/10^{10} RBC)	642, 548	3601, 3460
Red cell stromal ATPase	985	1500
(mμMoles Pi hydrolyzed/mg. stromal protein/hour)		
Red cell potassium (mEq./L. cells*)	42.5, 43.4	98, 95
Red cell sodium (mEq./L. cells*)	118, 114	23, 22.2
Potassium flux (mEq./L. cells*/hour)		
influx	3.0	3.7
outflux	3.4	3.7
net flux	—0.4	0.0
influx with ouabain (10^{-4}M)	0.8	0.6
ouabain inhibitable pump	2.2	3.1
Sodium flux (mEq./L. cells*/hour)		
influx	21.6	5.4
outflux	20.2	7.4
net flux	+1.4	-2.0
outflux with ouabain (10^{-4}M)	15.9	3.2
ouabain inhibitable pump	4.3	4.2

*Corrected to a liter of cells with an MCV of 85 cu. mμ.

a markedly accelerated pump rate. The leak rate in the older cells was much greater and was not sufficiently compensated for by the pump. The internal sodium, was, therefore, much higher in the old cells. These findings are summarized in Fig. 2 which relates internal sodium concentration to cation pumping. The young cells of patient E.S. pumped normally for their internal sodium concentration, but the old cells had a relatively deficient pump rate.

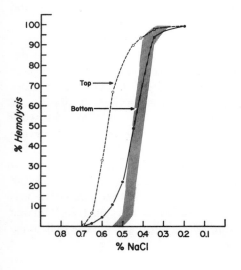

Fig. 1.—Osmotic fragility of erythrocytes separated after centrifugation. "Top" cells are the old cells and the "bottom" cells the young erythrocytes of E.S. (the father). (Reprinted with the permission of the authors and the New Eng. J. Med.)

Fig. 2.—Erythrocyte sodium and potassium pumping rates in response to the intracellular concentration of sodium. "Normal" values, taken from the literature and our own laboratory, are shown in the shaded area. The sodium data of Maizels are enclosed in circles. The sodium pump data of Parker and Hoffman are calculated from their measured potassium pump values on the basis of a 3:2 ratio. Note that the cells of R.Y., reported previously, pump at a rate far greater than expected from their internal sodium concentration and that the old cells of patient E.S. pump much less than expected. The values shown for E.S. are corrected to a cell volume of 85 cu. μ. (Reprinted with permission of the authors and the New Eng. J. Med.)

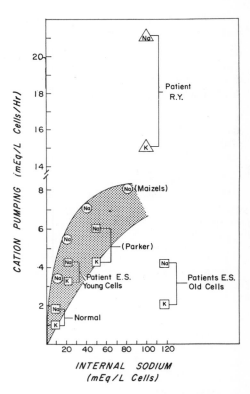

No explanation for decreased pumping in the old cells was found. ATPase activity was normal. Figure 2 also shows that the cells of the first patient described (R.Y.) had an extremely rapid rate of pumping, much higher than might have been expected from their internal sodium concentration. Finally, it was shown that the increased pumping present in E.S.'s cells was not associated either with a spheroidal shape or with lipid loss, lending additional evidence against a physiological link between pumping and lipid turnover.

Many questions were not answered by the studies of this fascinating family. Once again, the cause of increased permeability was not defined, nor was the basis of decreased pumping in the old cells discovered. Loss of a membrane carrier of cations in the old cells was postulated but proof for such an hypothesis has not been obtained. Of major importance was the lack of significant hemolysis in these patients. They were slightly anemic and ^{51}Cr survivals were slightly short with splenic sequestration detected during transfusion of the propositus' cells into a normal subject. But these cases showed that cation and water overload were well tolerated in human red cells, having little or no effect on their survival unless the overload was very extreme, as in the case of R.Y. At the other end of the spectrum, however, either membrane loss or cellular dehydration does create serious obstacles to red cell transit in the microcirculation, as illustrated by the studies of pyruvate kinase deficiency and sickle cell anemia to be described below.

Cells with Irregular Surfaces

Pyruvate kinase (PK) deficiency. PK deficiency was discovered to be an important cause of type II congenital nonspherocytic hemolytic anemia by Tanaka, Valentine and Miwa.[141] Since then, the clinical and enzymatic heterogeneity of this disease has become so apparent that it is difficult to make any broad generalizations. The morphology of the red cells in PK deficiency may be quite unimpressive, but in many cases, and particularly after splenectomy, severe erythrocyte spicule formation may be present. This bizarre deformation described in detail by Ponder[119] may occur in a variety of settings including postsplenectomy idiopathic thrombocytopenia or HS, cirrhosis with hyperlipemia, various "nonspherocytic" hemolytic anemias, uremia, abetalipoproteinemia, and hypothyroidism. In the initial case of sever spicule formation and PK deficiency described by Oski and Nathan and their co-workers,[104,112] the red cells of the patient remained irregular even when the cells were distended with hypotonic fluids. A striking abnormality in these cells was a markedly increased potassium flux with net potassium loss despite a high rate of potassium pumping. Glycolysis was nearly absent and ATP was unstable. The patient's nucleated erythroid precursors were not distorted by spicule formation, and a general hypothesis concerning these PK-deficient cells was advanced. First, the block in glycolysis was thought to be responsible for diminished ATP production. In addition, a membrane defect of unknown cause responsible for the morphologic disturbance and excessive potassium leak was proposed. The authors did not believe that the unstable ATP levels were directly responsible for spicule formation, despite the observations of Nakao and co-workers,[103] largely because normal red cell ghosts maintained their shape even when ATP was completely removed.[131] The authors also proposed that the accelerated potassium leak was only partially compensated for by accelerated pumping and that the increased rate of pumping further depleted cellular ATP. In fact, they found that when ouabain was exhibited to these cells, ATP was more stable. Since nucleated cells contain mitochondria and, therefore, maintain a higher rate of ATP synthesis, they were thought to be free of the metabolic and membrane defect. Indeed, as mentioned, the morphologic studies of the nucleated red cell precursors in this patient showed that they never were distorted by membrane spicules.

In further studies of a heterogeneous group of hemolytic anemias[105] characterized in part by spicule formation, ATP instability and increased red cell potassium flux, the authors showed that a block in glycolysis was not a necessary accompaniment of this triad. Furthermore, spicule formation itself, no matter how severe, was not necessarily associated with either abnormal membrane flux or even shortening of red cell survival. At this confusing moment, Keitt provided some important insight into the PK-deficient cell as a result of his studies of a patient with postsplenectomy reticulocytosis and PK deficiency.[72] He showed that the ATP in his patient's reticulocytes depended upon oxidative phosphorylation, as had been postulated for the nucleated red cells.[104,112] ATP fell rapidly in these reticulocytes when cyanide was present. Keitt concluded that red cells severely deficient in PK activity maintained their viability only during the reticulocyte stage of development when mitochondria were present. They were doomed to destruction at maturity. Reticulocytes, however, usually increased in the blood after splenectomy for PK deficiency,[14] suggesting that they might be selectively damaged by the spleen when it was present. Life span studies of reticulocytes and the whole blood of PK-deficient patients[107] showed that the young populations of cells which were rich in reticulocytes had the shortest survival and were destroyed in the liver upon escape from the spleen. On the other hand, metabolic studies utilizing mitochondrial substrates confirmed Keitt's observations. Reticulocytes of PK-deficient patients utilized fatty acids and amino acids; this metabolism was abolished by cyanide.[109]

The recent studies of Mentzer and his co-workers[87] have shed considerable light upon these seemingly paradoxical findings. Studying the metabolism of PK-deficient cells from patients with marked reticulocytosis following splenectomy, they noted cyanide-induced massive potassium loss, cell shrinkage and dehydration with resultant increased viscosity and decreased filterability. All of these events occurred concomitant with the rapid fall in cellular ATP. As suggested by the studies of Gardos,[42] Lepke and Passow,[77] Hoffman,[54] and Weed, LaCelle and Merrill,[149] the massive efflux of potassium and water induced by cyanide was retarded by EDTA, even though ATP loss continued unabated. These results provided evidence that reticulocytes in PK deficiency depended upon mitochondrial metabolism to maintain ATP and, secondarily, to maintain intracellular K^+. The EDTA effect suggested that the state of membrane Ca^{++} played a role in the K^+ loss. The dependence upon mitochondrial metabolism induced a vulnerability to hypoxia in the spleen. Indeed, exposure of these PK deficiency reticulocytes to a nitrogen atmosphere eventually produced effects very similar to those of cyanide. Furthermore, it was shown that the pO_2 at which oxygen consumption of PK deficiency reticuloyctes ceased was 40 mm. Hg, the pO_2 of venous blood. Thus, ion *loss* rather than ion *gain* provided the greatest challenge to the PK deficiency reticulocyte. Forced to depend upon mitochondria, it faced energy bankruptcy during prolonged exposure to the conditions of venous blood (in the spleen) because the mitochondria of adult reticulocytes could not function at venous pO_2 levels. As ATP fell and the relationship of Ca^{++} to the membrane was altered, K^+ left and with it went water. Mature PK-deficient cells which survived the splenic insult could circulate more freely because they produced enough ATP, despite their glycolytic block, to serve their relatively rudimentary needs.

This account of cell metabolism and permeability in PK deficiency emphasizes one important feature of red cell viability: the red cell may be even more sensitive to dehydration in vivo than it is to over hydration in vitro. The "desiccyte" has such a great increase in viscosity that it is unable to move through the strait passages in the microcirculation. Hence, it may be more vulnerable than the "hydrocyte" which retains its essential plastic characteristics.

Sickle cell disease. The shape change induced by hypoxia in sickle cells is well appreciated, as is the actual breakage induced in sickle cell membranes by repeated sickling and unsickling in vitro.[50] The morphology of blood smears in sickle cell disease reveals the result of intermittent sickling: fixed sickle forms, fragments and spherocytes. The spherocytes result from the loss of large pieces of membranes, as is described by Jensen and Lessin[68a] in this issue of SEMINARS.

Tosteson and co-workers[142] first commented on the metabolic consequences of the sickling phenomenon. They observed a marked increase in lactate production during deoxygenation of sickle cells, a change accompanied by excessive potassium loss. The observations of Mentzer and co-workers[86] have confirmed these findings. In fact, the rate of K^+ loss from sickle cells at various pO_2 levels has served as a sensitive index of liability to the sickling process of cells from patients with various sickle syndromes. The rate of hypoxia-induced K^+ loss also has provided a good index of clinical severity. In contrast to the findings in PK-deficient cells, the K^+ loss induced in deoxygenated sickle cells was not inhibited by EDTA. The K^+ loss in sickle cells, therefore, appeared to be independent of the state of membrane Ca^{++}. In addition, the K^+ loss was not balanced by the ingress of sodium. As a result, sickle cells became dehydrated during the sickling process, undoubtedly adding greatly to the increased viscosity induced by the shape change. The nature of the influence of the shape change on membrane permeability has not been established.

As might be expected, the K^+ loss induced in sickle cells by deoxygenation is accompanied by increased cation pumping, but the observed rise in glycolysis is not a result of the increment in pump activity since the glycolytic increment is only slightly inhibited by ouabain. The basis for the rise in glycolysis is not established, but its relationship to pH, the binding of glycolytic intermediates and of ATP, and the relative concentrations of certain glycolytic cofactors are under investigation because the deoxygenated sickle cell provides an important model for control mechanisms in red cell metabolism.

Irregular Cells with Irregular Ion Transport Defects

As mentioned above, cells with irregular surfaces are observed in several disease states, even when the life span of the red cell is normal. The acanthocytes described in the blood of patients with abetalipoproteinemia develop as the cells age,[135] as do the spicule cells observed in normal patients after splenectomy.[105] The cation concentrations and flux rates in these conditions are essentially normal. Patients with microangiopathic hemolytic anemia may have cells with enhanced K^+ flux, but the flux rates are usually appropriate for cell age. The spur cells observed in liver disease have behaved in a variable fashion in our limited experience. In one case, spur cell anemia plasma has been capable of inducing spicule formation and increased K^+ flux with net K^+ loss in normal cells.[105] In another case, currently under study by Shohet and Cooper, the patient's red cells appear to have a normal K^+ pump rate.

Red cells of adults with uremia frequently have spicules, whereas the cells

of children with uremia are singularly free of this morphologic abnormality.
Although we have been unable to detect a factor in uremic plasma capable
of inducing spicule formation in normal red cells, Cooper has designed conditions which provide good evidence that one or more heat labile and dialyz-
able plasma factors do induce the change.[25] Cation fluxes and ion contents are
normal in most uremic red cell samples, but in some patients with severe disease red cell sodium is increased and sodium efflux is decreased. Ouabain-
sensitive membrane ATPase is also decreased in such patients.[150] Whether or
not this alteration in pumping contributes to the shortening of the red cell life
span observed in severe uremia[70] is moot, but experience with other patients
with much higher red cell sodium and little or no hemolysis militates against
such a cause and effect relationship. The glycolysis of uremic or normal cells
is increased when it is measured in uremic plasma, buffered to pH 7.5,
probably because the phosphate concentration of uremic plasma is substan-
tially increased.[106] ATP is also usually increased in uremic cells. This observation can be explained as a combined result of increased glycolysis and normal
to low ATPase activity.[106,150]

Unstable Hemoglobins and Membrane Ion Transport

Exposure of normal or G-6-PD deficient red cells to oxidants such as
acetylphenylhydrazine creates hemoglobin precipitation and increased membrane cation permeability. Although the precipitation of the hemoglobin may
contribute to the increased membrane permeability by reacting with membrane thiols as it precipitates, it is more likely that direct attack on the
membrane by the oxidant chemical creates the enhanced permeability.

Study of the influence of hemoglobin precipitation on membrane function
has been assisted by the recognition that inherited unstable hemoglobins
create multiple small, or single large, inclusions in red cells. Therefore, studies
of inclusion-bearing cells can be made without exposure of the cells to damag-
ing chemicals. The unstable hemoglobins, such as Zurich, Köln and H, are
known to be associated with mild glutathione instability. Gabuzda and his
co-workers,[39] studying hemoglobin H, and Jacob and co-workers studying
hemoglobin Köln[64] have found that when the unstable hemoglobin precipitates, glutathione is bound to it, probably as a mixed disulfide. Jacob et al.
have found that K+ loss is enhanced in cells containing an unstable hemoglobin
when they are heated to 50°C. They offer the interesting suggestion that the
unstable hemoglobin may form mixed disulfides with membrane thiols when
it precipitates. This precipitation might lead to reorganization of proteins
in the membrane with a consequent increase in permeability. The red cells
of patients with hemolytic anemia due to precipitating hemoglobins, however,
have a young mean age and an increased turnover of ATP. The enhanced
potassium flux in these cells when they are heated may be related to rapid
ATP loss as well as to whatever membrane defects may be associated with
the unstable hemoglobin. We have not found it possible to separate the cells
of one patient with an as yet uncharacterized unstable hemoglobin into co-
horts with more or less precipitate. Inspection of blood samples from this

patient postsplenectomy reveals that nearly all of the cells have some precipitate. In the young cells, the precipitates are small and multiple; in the old cells only one large precipitate per cell is usually observed. It appears to be the product of coalescence of many small precipitates.

The thalassemia syndromes afford a special opportunity to investigate the effects of hemoglobin precipitation on membrane permeability because the development of alpha chain precipitates in homozygous beta thalassemia and of beta chain precipitates in hemoglobin H disease are influenced by red cell age.[108] Alpha chain precipitates form largely in the bone marrow and are less detectable in the cells with "balanced" chain production (the old or fetal hemoglobin-rich cells). Thus, in patients postsplenectomy, they are detected in the "young" red cell populations found at the top of a centrifuged column of cells and, in these same cells, potassium flux is increased. Hemoglobin H precipitates more slowly over the life span of the red cell. The inclusions of hemoglobin H, therefore, are found in the older cells of patients after splenectomy; at the bottom of a centrifuged column of cells. In these older cells containing inclusions, potassium flux is higher than it is in the young cells at the top of the centrifuged column. These data suggest that potassium flux does increase at the site of hemoglobin inclusions, but the basis of the increased ion flux is not at all clear. Rifkind[126] has shown that inclusion-bearing cells are liable to entrapment by the splenic littoral cells. In the spleen the red cells containing inclusions are attenuated, misshapen, and otherwise grossly injured. It seems reasonable to assume that the presence of inclusions leads to distortion and membrane injury of the cells that bear them as a result of their interaction with phagocytes within the reticuloendothelial (RE) system. The result is accelerated K^+ flux in the cells that have circulated with inclusions and have had injurious contact in vivo with RE cells.

At this time, it is not possible to decide whether inclusions directly injure red cell membranes by chemical means or if RE cells cause the damage by their interaction with inclusion-bearing cells. For reasons stated above, the degree of cation flux abnormality that is detected in inclusion-bearing cells is not great enough to be responsible for their shortened life span. It seems far more likely that the quietus delivered to an inclusion-bearing cell is mechanical. Its failure to pass through narrow endothelial channels causes a fatal confrontation with an avid phagocyte.

Minor Abnormalities of Red Cell Sodium Balance

In addition to their findings in uremic cells, Welt and his associates[151] have detected small elevations in red cell sodium in a heterogeneous group of severely ill patients whose diagnoses have included severe burns and terminal cancer. This defect is apparently due to a plasma inhibitor of erythrocyte ouabain sensitive ATPase activity. In fact, a high red cell sodium is taken as a poor prognostic sign by Welt and his co-workers. Their experience has been verified by studies of red cell sodium in experimental simian malaria. The plasma of severely infected monkeys inhibits the ouabain sensitive sodium efflux of normal red cells and the sodium content of the red cells of infected animals is considerably greater than that of normal monkeys.[34] The nature of the plasma factor which depresses ATPase activity has not yet been defined. An increased rate of red cell destruction is observed in all of these severely ill patients, but its relationship to the increased sodium concentration is not established.

During their studies, Balfe, Welt and their co-workers[2] found a family in which several members had decreased erythrocyte ATPase activity and increased red cell sodium concentrations. Hemolytic anemia was not present. Harvald and co-workers[52] had previously reported several cases of hemolytic anemia with decreased membrane ATPase activity, but electrolyte flux studies were not performed. Increased sodium pumping and slight elevation of red cell sodium were found in two cases of Bartter's syndrome; the red cell life span were normal.[40] A decrease in both ouabain sensitive and ethacrynic acid sensitive sodium efflux from the red cells of patients with cystic fibrosis has been reported.[3] In addition, a definite decrease was detected only in the ethacrynic acid-sensitive pump in the parents of affected patients. This observation has indicated a genetic marker for the carrier state which might be of considerable practical importance, as well as of theoretical interest.

SUMMARY

A review of red cell ion transport mechanisms and some of their relationships to red cell metabolism have been presented in this paper. Several human red cell disorders associated with abnormalities of cation transport have been described. Insufficient knowledge of the interaction of the reticuloendothelial system with abnormal red cells limits absolute assignments of pathophysiological roles to the observed abnormalities of ion transport. It appears, however, that the tendency to accumulate sodium and water in red cells is much less threatening to the life span of the red cell than is the tendency to lose potassium and water. The desiccated viscous red cell seems to be more rapidly removed from the circulation than is its bloated counterpart.

ACKNOWLEDGMENT

One of the studies of cell water was kindly performed by Dr. Joseph Hoffman, Yale, University School of Medicine, New Haven, Conn.

REFERENCES

1. Asakura, T., Sato, Y., Minakami, S., and Yoshikawa, H.: Effect of deoxygenation of intracellular hemoglobin on red cell glycolysis. J. Biochem. (Tokyo) 59:524, 1966.

2. Balfe, J. W., Cole, C., Smith, E. K. M., Graham, J. B., and Welt, L. G.: A hereditary sodium transport defect in the human red blood cell. J. Clin. Invest. 47:4a, 1968.

3. —, —, and Welt, L. G.: Red cell transport defect in patients with cystic fibrosis and in their parents. Science 162:689, 1968.

4. Bangham, A. D., Standish, M. M., and Watkins, J. C.: Diffusion of univalent ions across the lamellae of swollen phospholipids. J. Molec. Biol. 13:238, 1965.

5. Benesch, R., and Benesch, R. E.: The effect of organic phosphates from the human erythrocyte on the allosteric properties of hemoglobin. Biochem. Biophys. Res. Commun. 26:162, 1967.

6. Benesch, R. E., Benesch, R., and Yu, C. I.: The oxygenation of hemoglobin in the

presence of 2,3-diphosphoglycerate. Effect of temperature, pH, ionic strength, and hemoglobin concentration. Biochemistry (Wash.) 8:2567, 1969.

7. Berendes, M.: The proportion of reticulocytes in the erythrocytes of the spleen as compared with those of circulating blood, with special reference to hemolytic states. Blood 14:558, 1959.

8. Bernstein, R. E.: Alterations in metabolic energetics and cation transport during aging of red cells. J. Clin. Invest. 38:1572, 1959.

9. Bertles, J. F.: Sodium transport across the surface membrane of red blood cells in hereditary spherocytosis. J. Clin. Invest. 36:816, 1957.

10. Blum, S. F., and Oski, F. A.: Red cell metabolism in the newborn infant. IV. Transmembrane potassium flux. Pediatrics 43:396, 1969.

11. —, Shohet, S. B., Nathan, D. G., and Gardner, F. H.: The effect of amphotericin B on erythrocyte membrane cation permeability: its relation to in vivo erythrocyte survival. J. Lab. Clin. Med. 73:980, 1969.

12. Bolingbroke, V., and Maizels, M.: Calcium ions and the permeability of human erythrocytes. J. Physiol. (London) 149:563, 1959.

13. Bosch, H. van den, and van Deenen, L. L.: The formation of isomeric lysolecithins. Biochim. Biophys. Acta 84:234, 1964.

14. Bowman, H. S., and Procopio, F.: Hereditary non-spherocytic hemolytic anemia of pyruvate kinase deficient type. Ann. Intern. Med. 58:567, 1963.

15. Brewer, G. J., Eaton, J. W., Beck, C. C., Feitter, L., and Shreffer, D. C.: Sodium-potassium stimulated ATPase activity of mammalian hemolysates: clinical observations and dominance of ATPase deficiency in the potassium polymorphism of sheep. J. Lab. Clin. Med. 71:744, 1968.

16. Chanutin, A. A., and Curnish, R. R.: Effect of organic and inorganic phosphates on the oxygen equilibrium of human erythrocytes. Arch. Biochem. 121:96, 1967.

17. Chapman, R. G., and Schaumberg, L.: Glycolysis and glycolytic enzyme activity of aging red cells in man. Changes in hexokinase, aldolase, glyceraldehyde-3-phosphate dehydrogenase, pyruvate kinase and glutamic-oxalacetic transaminase. Brit. J. Haemat. 13:665, 1967.

18. —, and McDonald, L. L.: Red cell life span after splenectomy in hereditary spherocytosis. J. Clin. Invest. 47:2263, 1968.

19. Clark, W. M.: Topics in Physical Chemistry (ed. 2). Baltimore, Williams & Wilkins, 1952, p. 349.

20. Cohen, G., and Hochstein, P.: Glucose-6-phosphate dehydrogenase and detoxification of hydrogen peroxide in human erythrocytes. Science 134:1756, 1961.

21. —, and —: Glutathione peroxidase: the primary agent for the elimination of hydrogen peroxide in erythrocytes. Biochemistry (Wash.) 2:1420, 1963.

22. —, and —: Generation of hydrogen peroxide in erythrocytes by hemolytic agents. Biochemistry (Wash.) 3:895, 1964.

23. Cooper, R. A., and Jandl, J. H.: The role of membrane lipids in the survival of red cells in hereditary spherocytosis. J. Clin. Invest. 48:736, 1969.

24. —, and —: The selective and conjoint loss of red cell lipids. J. Clin. Invest. 48:906, 1969.

25. —: The pathogenesis of Burr cells in uremia. J. Clin. Invest. 49:22a, 1970.

25a. —: Lipids of human red cell membrane: normal composition and variability in disease. Seminars Hemat. 7:296, 1970.

26. Danielli, J. F., and Davson, H.: A contribution to the theory of permeability of thin films. J. Cell. Physiol. 5:495, 1935.

27. Watkins, W. M.: Blood-group substances. In Davis, B. D. and Warren, L. (Eds.): Specificity of Cell Surfaces. Englewood Cliffs, N. J., Prentice Hall, 1967, p. 257.

28. Davson, H., and Danielli, J. F.: The Permeability of Natural Membranes. London, Cambridge University Press, 1952.

29. Dawson, R. M. C.: Metabolism of animal phospholipids and their turnover in cell membranes. Assays Biochem. 2:69, 1966.

30. Desforges, J. F.: Erythrocyte metabolism in hemolysis. New Eng. J. Med. 273:1310, 1965.

31. Donnan, F. G., and Guggenheim, E. A.: Die Genaue thermodynamik der Membrangleichagewichte. Z. Physik. Chem. A-162:346, 1932.

32. Dormany, T. L.: Biological rancidification. Lancet 2:684, 1969.

33. Ducrou, W., and Kimber, R. J.: Stomatocytes, haemolytic anaemia and abdominal pain in Mediterrean migrants. Med. J. Aust. 2:1087, 1969.

34. Dunn, M. J.: Alterations of red blood

cell sodium transport during malarial infection. J. Clin. Invest. 48:674, 1969.

35. Eckel, R. E., Rizzo, S. C., Lodish, H., and Berggren, A. B.: Potassium transport and control of glycolysis in human erythrocytes. Amer. J. Physiol. 210:737, 1966.

36. Emerson, C. P., Jr., Shen, S. C., Ham, T. H., Fleming, E. M., and Castle, W. B.: Studies on the destruction of red blood cells. IX. Quantitative methods for determining the osmotic and mechanical fragility of red cells in the peripheral blood and splenic pulp; the mechanism of increased hemolysis in hereditary spherocytosis (congenital hemolytic jaundice) as related to the function of the spleen. Arch. Intern. Med. (Chicago) 97:1, 1956.

37. Feig, S. A., Shohet, S. B., and Nathan, D. G.: Fluoride inhibition of red cell (RBC) membrane function. Clin. Res. 18: 403, 1970.

38. Ferber, E., Krüger, J., Munder, P. G., Kohlschütter, A., and Fischer, H.: Acyltransferase- und Lysophospholipase-Aktivität in Membranen von Erythrocyten während der Alterung in vivo und in vitro. In Deutsch, E., Gerlach, E., and Moser, K. (Eds.): Stoffwechsel und Membranpermeabilität von Erythrocyten und Thrombocyten. Stuttgart, Georg Thieme, Verlag, 1968, p. 393.

39. Gabuzda, T. G., Laforet, M. T., and Gardner, F. H.: Oxidative precipitation of hemoglobin H and its relation to reduced glutathione. J. Lab. Clin. Med. 70:581, 1967.

40. Gall, G., Haddow, J., Vaitukaitas, S., and Klein, R.: Sodium flux across the red cell membrane in patients with Bartter's syndrome and with hyperal dosteronism. Seventy-ninth Annual Meeting of the Amer. Ped. Soc., 1969 (abstract).

41. Gallai-Hatchard, J. J., and Thompson, R. H.: Phospholipase-A activity of mammalian tissues. Biochim. Biophys. Acta 98: 128, 1965.

42. Gardos, G.: The function of calcium in the regulation of ion transport In Kleinzeller, A., and Kotyka, A. (Eds.): Membrane Transport and Metabolism. New York, Academic Press, 1960, p. 553.

43. Green, H., Barrow, P., and Goldberg, B.: Effect of antibody and complement on permeability control in ascites tumor cells and erythrocytes. J. Exp. Med. 110:699, 1959.

44. —, and Goldberg, B.: The action of antibody and complement on mammalian cells. Ann. N. Y. Acad. Sci. 87:352, 1960.

45. Grimes, A. J., Meisler, A., and Dacie, J. V.: Hereditary non-spherocytic haemolytic anaemia. A study of red-cell carbohydrate metabolism in twelve cases of pyruvate-kinase deficiency. Brit. J. Haemat. 10: 403, 1964.

46. Guest, G. M.: Osmometric behavior of normal and abnormal human erythrocytes. Blood 3:541, 1948.

47. Ham, T. H., and Castle, W. B.: Relation of increased hypotonic fragility and of erythrostasis to mechanism of hemolysis in certain anemias. Trans. Ass. Amer. Physicians 55:127, 1940.

48. Harris, E. J., and Prankerd, T. A.: The rate of sodium extrusion from human erythrocytes. J. Physiol. (London) 121:470, 1953.

49. Harris, I. M., McAlister, J. M., and Prankerd, T. A.: The relationship of abnormal red cells to the normal spleen. Clin. Sci. 16:223, 1957.

50. Harris, J. W., Brewster, H. H., Ham, T. H., and Castle, W. B.: Studies on the destruction of red blood cells. X. The biophysics and biology of sickle-cell disease. Arch. Intern. Med. (Chicago) 97:145, 1956.

51. Harrison, D. G., and Long, C.: The calcium content of human erythrocytes. J. Physiol. (London) 199:367, 1968.

52. Harvald, B., Hanel, K. H., Squires, R., and Trap-Jensen, J.: Adenosine-triphosphatase deficiency in patients with non-spherocytic haemolytic anaemia. Lancet 2:18, 1964.

53. Hoffman, J. F.: Cation transport and structure of the red cell plasma membrane. Circulation 26:1201, 1962.

54. —: The red cell membrane and the transport of sodium and potassium. Amer. J. Med. 41:666, 1966.

55. Hokin, L. E., and Hokin, M. R.: Diglyceride kinase and phosphatidic acid phosphatase in erythrocyte membranes. Nature (London) 189:836, 1961.

56. Engström, A., and Finean, J. B.: Biological Ultrastructure. New York, Academic Press, 1958, p. 237.

57. Jacob, H. S., and Jandl, J. H.: Effects of sulfhydryl inhibition on red blood cells. I. Mechanism of hemolysis. J. Clin. Invest. 41:779, 1962.

58. —, and —: Increased cell membrane permeability in the pathogenesis of heredi-

tary spherocytosis. J. Clin. Invest. 43:1704, 1964.

59. —, Ingbar, S. H., and Jandl, J. H.: Oxidative hemolysis and erythrocyte metabolism in hereditary acatalasia. J. Clin. Invest. 44:1187, 1965.

60. —: Annotation: Dysfunction of the red cell membrane in hereditary spherocytosis. Brit. J. Haemat. 14:99, 1968.

61. —, and Jandl, J. H.: Effects of sulfhydryl inhibition on red blood cells. III. Glutathione in the regulation of the hexose monophosphate pathway. J. Biol. Chem. 241:4243, 1966.

62. —, and Karnovsky, M. L.: Concomitant alterations of sodium flux and membrane phospholipid metabolism in red blood cells: studies in hereditary spherocytosis. J. Clin. Invest. 46:173, 1967.

63. —, and Lux, S. E., IV: Degradation of membrane phospholipids and thiols in peroxide hemolysis: studies in vitamin E deficiency. Blood 32:549, 1968.

64. —, Brain, M. C., and Dacie, J. V.: Altered sulfhydryl reactivity of hemoglobins and red blood cell membranes in congenital Heinz body hemolytic anemia. J. Clin. Invest. 47:2664, 1968.

65. Jandl, J. H.: The agglutination and sequestration of immature red cells. J. Lab. Clin. Med. 55:663, 1960.

66. —, Simmons, R. L., and Castle, W. B.: Red cell filtration and the pathogenesis of certain hemolytic anemias. Blood 18:133, 1961.

67. —: Leaky red cells. Blood 26:367, 1965.

68. Jandl, J. H., and Aster, R. H.: Increased splenic pooling and the pathogenesis of hypersplenism. Amer. J. Med. Sci. 253:383, 1967.

68a. Jensen, W. N., and Lessin, L.: Membrane alterations associated with hemoglobinopathies. Seminars Hemat. 7:409, 1970.

69. Joyce, C. R. B.: Uptake of potassium and sodium by parts of packed human blood cell column. Quart. J. Exp. Physiol. 43:299, 1958.

70. Kaye, M.: The anemia associated with renal disease. J. Lab. Clin. Med. 52:83, 1958.

71. Kedem, O., and Katchalsky, A.: A physical interpretation of the phenomenological coefficients of membrane permeability. J. Gen. Physiol. 45:143, 1961.

72. Keitt, A. S.: Pyruvate kinase deficiency and related disorders of red cell glycolysis. Amer. J. Med. 41:762, 1966.

73. Kinsky, S. C., Avruch, J., Permutt, M., and Rogers, H. B.: The lytic effect of polyene antifungal antibiotics on mammalian erythrocytes. Biochem. Biophys. Res. Commun. 9:503, 1962.

74. —, Luse, S. A., and van Deenen, L. L.: Interaction of polyene antibiotics with natural and artificial membrane systems. Fed. Proc. 25:1503, 1966.

75. Lane, D., and Castle, W. B.: Personal communication.

76. Langley, G. R., and Axell, M.: Changes in erythrocyte membrane and autohaemolysis during in vitro incubation. Brit. J. Haemat. 14:593, 1968.

77. Lepke, S., and Passow, H.: Effects of fluoride on potassium and sodium permeability of the erythrocyte membrane. J. Gen. Physiol. 51:365S, 1968.

78. Lloveras, J., Douste-Blazy, L., and Valdiguie, P.: Mode of action of splenic phospholipase. C. R. Acad. Sci. (D) (Paris) 256:1861, 1963.

79. LoBuglio, A. F., Cotran, R. S., and Jandl, J. H.: Red cells coated with immunoglobulin G: binding and sphering by mononuclear cells in man. Science 158:1582, 1967.

80. Lock, S. P., Smith, R. S., and Hardisty, R. M.: Stomatocytosis: a hereditary red cell anomaly associated with haemolytic anaemia. Brit. J. Haemat. 1:303, 1961.

81. Lowry, O.: Phosphofructokinase. In Chance, B., Estabrook, R. W., and Williamson, J. R. (Eds.): Control of Energy Metabolism. New York, Academic Press, 1965, p. 63.

82. Lubin, B. H., and Shohet, S. B.: Alterations in membrane fatty acid (FA) turnover in vitamin E deficient erythrocytes (E-RBC) during exposure to hydrogen peroxide. Presented to Soc. Ped. Res., 1970.

83. Maizels, M.: Anion and cation contents of normal and anaemic bloods. Biochem. J. 30:821, 1936.

84. —: Calcium ions and the permeability of human red cells. Nature (London) 184:366, 1959.

85. Meadow, S. R.: Stomatocytosis. Proc. Roy. Soc. Med. 60:13, 1967.

86. Mentzer, W. C., August, C. S., and Nathan, D. G.: The effects of androgen administration in sickle cell anemia. Pediat.

Res. 3:378, 1969.

87. —, Baehner, R. L., Schmidt-Schoenbein, H., and Nathan, D. G.: Metabolic vulnerability of reticulocytes in pyruvate kinase (PK) deficiency. Blood 34:861, 1969.

88. Miller, G., Townes, P. L., and MacWhinney, J. B.: A new congenital hemolytic anemia with deformed erythrocytes ("stomatocytes") and remarkable susceptibility of erythrocytes to cold hemolysis in vitro. I. Clinical and hematologic studies. Pediatrics 35:906, 1965.

89. Mills, G. C.: The purification and properties of glutathione peroxidase of erythrocytes. J. Biol. Chem. 234:502, 1959.

90. —: Effects of pH on erythrocyte metabolism. Arch. Biochem. 117:487, 1966.

91. Minakami, S., Kakinuma, K., and Yoshikawa, H.: The control of erythrocyte glycolysis by active cation transport. Biochim. Biophys. Acta 90:434, 1964.

92. —, and Yoshikawa, H.: Inorganic phosphate and erythrocyte glycolysis. Biochim. Biophys. Acta 99:175, 1965.

93. —, and —: Studies on erythrocyte glycolysis. III. The effects of active cation transport, pH, and inorganic phosphate concentration on erythrocyte glycolysis. J. Biochem. (Tokyo) 59:145, 1966.

94. Miróevová, L., Brabec, V., and Palek, J.: Adenosintriphosphatase activity in hereditary spherocytosis. Blut 15:149, 1967.

95. Mohler, D. N.: Adenosine triphosphate metabolism in hereditary spherocytosis. J. Clin. Invest. 44:1417, 1965.

96. Mollison, P. L., and Young, I. M.: Failure of in vitro tests as a guide to value to stored blood. Brit. Med. J. 2:797, 1941.

97. —, and —: In vivo survival in the human subject of transfused erythrocytes after storage in various preservative solutions. Quart. J. Exp. Physiol. 31:359, 1942.

98. Muir-Jackson, J., and Knight, D.: Stomatocytosis in migrants of Mediterranean origin. Med. J. Aust. 1:939, 1969.

99. Mulder, E., VanDenBerg, J. W. O., and van Deenen, L. L. M.: Metabolism of red-cell lipids. II. Conversion of lysophosphoglycerides. Biochim. Biophys. Acta 106:118, 1965.

100. Murphy, J. R.: Erythrocyte metabolism. III. Relationship of energy metabolism and serum factors to the osmotic fragility following incubation. J. Lab. Clin. Med. 60:86, 1962.

101. —: Erythrocyte metabolism. V. Active cation transport and glycolysis. J. Lab. Clin. Med. 61:567, 1963.

102. —: The influence of pH and temperature on some physical properties of normal erythrocytes and erythrocytes from patients with hereditary spherocytosis. J. Lab. Clin. Med. 69:758, 1967.

103. Nakao, M., Nakao, T., and Yamazoe, S.: Adenosine triphosphate and maintenance of shape of the human red cells. Nature (London) 187:945, 1960.

104. Nathan, D. G., Oski, F. A., Sidel, V. W., and Diamond, L. K.: Extreme hemolysis and red cell distortion in erythrocyte pyruvate kinase deficiency. II. Measurements of erythrocyte glucose consumption potassium flux and adenosine triphosphate stability. New Eng. J. Med. 272:118, 1965.

105. —, Oski, F. A., Sidel, V. W., Gardner, F. H., and Diamond, L. K.: Studies of erythrocyte spicule formation and haemolytic anaemia. Brit. J. Haemat. 12:385, 1966.

106. —, Beck, L. H., Hampers, C. L., and Merrill, J. P.: Erythrocyte production and metabolism in anephric and uremic men. Ann. N. Y. Acad. Sci. 149:539, 1968.

107. —, Oski, F. A., Miller, D. R., and Gardner, F. H.: Life span and organ sequestration of the red cells in pyruvate kinase deficiency. New Eng. J. Med. 278:73, 1968.

108. —, Stossel, T. B., Gunn, R. B., Zarkowsky, H. S., and Laforet, M. T.: Influence of hemoglobin precipitation on erythrocyte metabolism in alpha and beta thalassemia. J. Clin. Invest. 48:33, 1969.

109. —: Unpublished observations.

110. Norman, J. G.: Stomatocytosis in migrants of Mediterranean origin. Med. J. Aust. 1:315, 1969.

111. Olmstead, E. G.: Efflux and influx of erythrocyte water. J. Gen. Physiol. 44:227, 1960.

112. Oski, F. A., Nathan, D. G., Sidel, V. W., and Diamond, L. K.: Extreme hemolysis and red cell distortion in erythrocyte pyruvate kinase deficiency. I. Morphology, erythrokinetics and family enzyme studies. New Eng. J. Med. 270:1023, 1964.

113. —, Naiman, J. L., Blum, S. F., Zarkowsky, H. S., Whaun, J., Shohet, S. B., Green, A., and Nathan, D. G.: Congenital hemolytic anemia with high-sodium, low-potassium red cells. Studies of three generations of a family with a new variant. New Eng. J. Med. 280:909, 1969.

114. Parker, J. C., and Hoffman, J. F.: The role of membrane phosphoglycerate kinase in the control of glycolytic rate by active cation transport in human red blood cells. J. Gen. Physiol. 50:893, 1967.

115. Parpart, A. K., Ballentine, R., and Barron, E. S. G. (Eds.): Modern Trends in Physiology and Biochemistry. New York, Academic Press, 1957, p. 135.

116. Passow, H.: Ion and water permeability of the red blood cell. In Bishop, C., and Surgenor, D. M. (Eds.): The Red Blood Cell. New York, Academic Press, 1964, p. 71.

117. Perutz, M. F., Muirhead, H., Mazzarella, L., Crowther, R. A. Greer, J., and Kilmartin, J. V.: Identification of residues responsible for the alkaline Bohr effect in haemoglobin. Nature (London) 222:1240, 1969.

118. Peters, J. C., Rowland, M., Israels, L. G., and Zipursky, A.: Erythrocyte sodium transport in hereditary elliptocytosis. Canad. J. Physiol. Pharmacol. 44:817, 1966.

119. Ponder, E.: Hemolysis and Related Phenomena. New York, Grune & Stratton, 1948.

120. Post, R. L., Merritt, G. R., Kinsolving, C. R., and Albright, C. D.: Membrane adenosine triphosphatase as a participant in the active transport of sodium and potassium in the human erythrocyte. J. Biol. Chem. 235:1796, 1960.

121. Prankerd, T. A. J.: Studies on the pathogenesis of haemolysis in hereditary spherocytosis. Quart. J. Med. 29:199, 1960.

122. —: The Red Cell. England, Blackwell Scientific Publications, 1961, p. 63.

123. Ibid. p. 97.

124. Reed, C. F., and Swisher, S. N.: Erythrocyte lipid loss in hereditary spherocytosis. J. Clin. Invest. 45:777, 1966.

125. —: Phospholipid exchange between plasma and erythrocytes in man and dog. J. Clin. Invest. 47:749, 1968.

126. Rifkind, R. A.: Destruction of injured red cells in vivo. Amer. J. Med. 41:711, 1966.

127. Rosenberg, S. A., and Guidotti, G.: Fractionation of the protein components of human erythrocyte membranes. J. Biol. Chem. 244:5118, 1969.

128. Rowe, C. E.: The phospholipids of human-blood plasma and their exchange with the cells. Biochem. J. 76:471, 1960.

129. Sass, M. D., Caruso, C. J., and Axelrod, D. R.: Accumulation of methylene blue by metabolizing erythrocytes. J. Lab. Clin. Med. 69:447, 1967.

130. —, Caruso, J., and Farhangi, M.: TPNH-methemoglobin reductase deficiency: a new red cell enzyme defect. J. Lab. Clin. Med. 70:760, 1967.

131. Schrier, S. L.: Studies of the metabolism of human erythrocyte membranes. J. Clin. Invest. 42:756, 1963.

132. —: ATP synthesis in human erythrocyte membranes. Biochim. Biophys. Acta 135:591, 1967.

133. —, Moore, L. D., and Chiapella, A. P.: Inhibition of human erythrocyte membrane-mediated ATP synthesis by anti-D antibody. Amer. J. Med. Sci. 256:340, 1968.

134. Sha'afi, R. I., Rich, G. T., Sidel, V. W., Bossert, W., and Solomon, A. K.: Effect of unstirred layer on human red cell permeability. J. Gen Physiol. 50:1377, 1967.

135. Simon, E. R., and Ways, P.: Incubation hemolysis and red cell metabolism in acanthocytosis. J. Clin. Invest. 43:1311, 1964.

136. Shohet, S. B.: Increased ^{14}C fatty acid (FA) incorporation into erythrocyte membrane phospholipids in hemolytic states. Clin. Res. 16:314, 1968.

137. Solomon, A. K., Gill, T. J. III, and Gold, G. L.: The kinetics of cardiac glycoside inhibition of potassium transport in human erythrocytes. J. Gen. Physiol. 40:327, 1956.

138. —: Red cell membrane structure and ion transport. J. Gen. Physiol. 43 (suppl. 1: 1, 1960.

139. —: Measurements of the equivalent pore radius in cell membranes. In Kleinzeller, A., and Kotyk, A. (Eds.): Membrane transport and metabolism. New York, Academic Press, 1960, p. 94.

140. —: Characterization of biological membranes by equivalent pores. J. Gen. Physiol. 51:335S, 1968.

141. Tanaka, K. R., Valentine, W. W., and Miwa, S.: Pyruvate kinase (PK) deficiency hereditary non-spherocytic hemolytic anemia. Blood 19:267, 1962.

142. Tosteson, D. C., Carlsen, E., and Dunham, E. T.: The effects of sickling on ion transport. I. Effect of sickling on potassium transport. J. Gen. Physiol. 39:31, 1955.

143. —: Halide transport in red blood cells. Acta Physiol. Scand. 46:19, 1959.

144. van Deenen, L. L. M.: Phospholipids

and biomembranes. *In* Holman, R. T. (Ed.): Progress in the Chemistry of Fats and Other Lipids, VIII, Part 1. New York, Pergamon Press, 1965, p. 64.

145. Van Slyke, D. D., Wu, H., and McLean, F. C.: Studies of gas and electrolyte equilibria in the blood. V. Factors controlling the electrolyte and water distribution in the blood. J. Biol. Chem. 56:765, 1923.

146. Watkins, W. M.: Blood group substances. Science 152:172, 1966.

147. Weed, R. I.: Effects of primaquine on the red blood cell membrane. II. Potassium ion permeability in glucose-6-phosphate dehydrogenase deficient erythrocytes. J. Clin. Invest. 40:140, 1961.

148. —, and Bowdler, A. J.: Metabolic dependence of the critical hemolytic volume of human erythrocytes: relationship to osmotic fragility and autohemolysis in hereditary spherocytosis and normal red cells. J. Clin. Invest. 45:1137, 1966.

149. —, LaCelle, P. L., and Merrill, E. W.: Metabolic dependence of red cell deformability. J. Clin. Invest. 48:795, 1969.

150. Welt, L. G., Sachs, J. R., and McManus, T. J.: An ion transport defect in erythrocytes from uremic patients. Trans. Ass. Amer. Physicians 77:169, 1964.

151. —, Smith, E. K., Dunn, M. J., Czerwinski, A., Proctor, H., Cole, C., Balfe, J. W., and Gitleman, H. J.: Membrane transport defect: the sick cell. Trans. Ass. Amer. Physicians 80:217, 1967.

152. Wennberg, E., and Weiss, L.: The structure of the spleen and hemolysis. Ann. Rev. Med. 20:29, 1969.

153: Whittam, R.: Transport and Diffusion in Red Blood Cells. Baltimore, Williams & Wilkins Co., 1964, p. 41.

154. *Ibid.* p. 103.

155. *Ibid.* p. 106.

156. *Ibid.* p. 109.

157. *Ibid.* p. 118.

158. *Ibid.* p. 130.

159. *Ibid.* p. 150.

160. Wiley, J. S.: Dominant inheritance of the sodium leak in hereditary spherocytic red cells and a comparison of sodium leak with red cell survival. Aust. Ann. Med. 17: 177, 1968.

161. Wiley, J.: Inheritance of an increased sodium pump in human red cells. Nature (London) 221:1222, 1969.

162. Zarkowsky, H. S., Oski, F. A., Sha'afi, R., Shohet, S. B., and Nathan, D. G.: Congenital hemolytic anemia with high sodium, low potassium red cells. I. Studies of membrane permeability. New Eng. J. Med. 278:573, 1968.

163. —, and Nathan, D. G.: Influence of erythrocyte membrane ATPase on the metabolism of hemolysates. J. Lab. Clin. Med. 76:231, 1970.

164. —, and Nathan, D. G.: Unpublished data.

Membrane Alterations Associated with Hemoglobinopathies

By WALLACE N. JENSEN AND LAWRENCE S. LESSIN

S INCE THE DESCRIPTION OF SICKLE CELL ANEMIA as a molecular disease[24] caused by a genetically determined alteration of the hemoglobin molecule, over 100 other molecular variants of human hemoglobin have been described.[25] The abnormalities are due to substitutions and/or deletions of one or more amino acid in the globin chains and may be classified into four categories:

(1) Those which cause no detectable functional or pathologic change of the hemoglobin molecule or the red cell (for example, HbJ Oxford).

(2) Those with abnormal oxygen association characteristics in which the oxygen affinity is either increased (Hb Chesapeake, Hb Yakima, Hb Seattle, etc.) or decreased (Hb Kansas) and may, respectively, cause increased or decreased red cell mass.

(3) Those associated with hemoglobin instability and which are most commonly due to substitutions of amino acids which occupy the heme cavity of the globin moiety of the molecule. Certain amino acid substitutions in this portion of either the alpha- or beta-globin chains enhance methemoglobin production and, at times, cause denaturation and particulate precipitation of hemoglobin (Heinz bodies).

(4) Those in which polar and nonpolar amino acid substitutions cause alteration of the external surface of the hemoglobin which in turn results in molecular polymerization and the formation of a paracrystalline (HbC) or a helical type (HbS) protein polymer.

Of the hemoglobinopathies which have been adequately studied, it may be generalized that shortening of red cell life span occurs only when there is a pathologic molecular interaction between hemoglobin aggregates, polymers or crystals of the denatured or deoxygenated protein and the cell membrane. There is evidence which indicates that hemoglobin molecules are tightly bound to and may have a structural role in the red cell membrane.[29] Hemoglobins, exhibiting abnormal molecular interactions, might therefore be expected to alter membrane function and structure. Such alterations are demonstrable by examination of cell ultrastructure with various methods and are seen in the accompanying photomicrographs.

Hemolysis, secondary to membrane dysfunction in a cell which contains abnormal hemoglobins, may occur by at least two mechanisms:

From The George Washington University Medical Center, Washington, D.C.

Supported in part by USPHS Grant 7R01-AM-14324-01.

WALLACE N. JENSEN, M.D.: *Professor and Chairman, Department of Medicine, The George Washington University, Washington, D.C.* LAWRENCE S. LESSIN, M.D.: *Division of Hematology, Department of Medicine, The George Washington University, Washington, D.C.*

161

(1) Fragmentation with symmetrical loss of membrane lipids leading to formation of microspherocytes and splenic or reticuloendothelial cell erythrophagocytosis of the resultant rigid or deformed cell.[33,34]

(2) Alteration in cation permeability of the red cell membrane due to interaction of precipitated or aggregated denatured hemoglobin or globin with the cell membrane.[11] The increased cation permeability eventually causes osmotic lysis of the cell.

The Unstable Hemoglobins: Heinz Bodies and Membrane Deformation

There are undoubtedly different types of denatured hemoglobin aggregates which are collectively termed Heinz bodies.[13,16]

In alpha thalassemia (Hb H disease), the unstable protein complexes (Heinz bodies) are formed from the excess beta chains which result from a relative suppression of synthesis of alpha chains.[28] Morphologically similar, denatured hemoglobin precipitates are present in red cells with enzymatic defects of the pentosephosphate pathway,[6] or following exposure to potent oxidizing agents, drugs or toxins.[30] Unstable hemoglobins are usually attributed to mutations which dictate change in an amino acid residue in close contact with a heme group, or which cause an alteration in the conformation of the heme cavity.[5]

Morphologically, hemoglobin instability may be manifest by the spontaneous development of Heinz bodies in erythrocytes, reticulocytes, or erythroblasts[5] (Bessis, unpublished data). In phase contrast microscopy, Heinz bodies appear as dense, irregular inclusions ranging in size from the limits of optical resolution to 2 μ in diameter which are commonly fixed to the cell membrane (Fig. 1A). The membrane attachment is more apparent with use of the interference microscope where the Heinz bodies appear as "bumps" or "pits" on the red cell surface (Fig. 1B). At higher magnification, the scanning electron microscope (SEM) permits visualization of surface membrane irregularities in greater detail. Aggregates of denatured hemoglobin fixed to the cell membrane are very insoluble as is inferred by the SEM photographs in Fig. 2 A–F. With conventional light microscopy, Heinz bodies are only recognized with supravital staining (new methylene blue, brilliant cresyl blue) and appear as multiple irregular azurophilic intracellular aggregates (Fig. 1C). Transmission electron microscopy (TEM) shows that Heinz body formation begins with the microaggregation of denatured hemoglobin, appearing as centrally located osmophilic electron dense bodies[26] (500–600 Å in diameter) (Fig. 1D). With progression of Heinz body formation, the bodies coalesce to form larger osmophilic masses which migrate toward and become fixed to the red cell membrane (Fig. 1E). In reticulocytes or erythroblasts the denatured hemoglobin may undergo coprecipitation with organelles, including ribosomes, mitochondria, siderosomes and cytoplasmic ferritin (Bessis, unpublished data).

The fixation of the Heinz bodies to the cell membrane produces regional membrane alterations and changes in osmotic and mechanical fragility.[5] The observation that cells with Heinz bodies are more numerous after splenectomy indicates the importance of sequestration of the abnormal cells.[27,30] The cells with Heinz bodies do not easily traverse the 2.5 to 3.0 μ

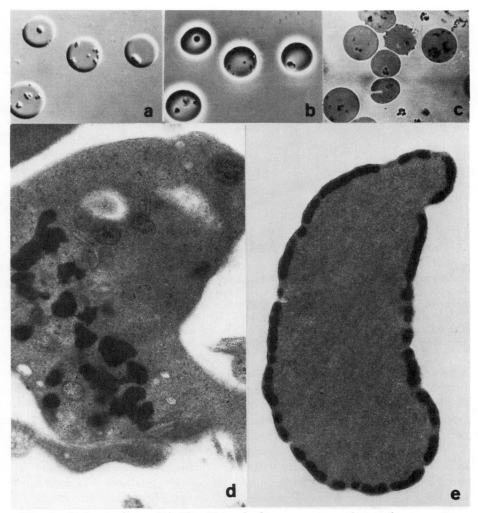

Fig. 1.—Heinz bodies. (A) The Heinz bodies are seen with interference micro-scopy and may be single, or multiple and range up to 0.2μ in size. Fixation to the membrane is evident and they appear as projections from the surface of the cell. (B) The same Heinz bodies as seen in Fig. 1 A viewed with phase microscopy. The inclusions are dense, well defined and are coalescent near the center of the cell. (C) The same Heinz bodies seen in Fig. 1 A and 1 B, but treated with new methylene blue stain now appear as multiple irregular inclusions with membrane fixation. (D) An electron microscopic section of a reticulocyte with Heinz bodies. Note the multiple intracellular electron dense bodies which tend to fuse, initially at the center of the cell. Coprecipitation of the denatured hemoglobin with mitochondria, ribosomes, siderosomes and cytoplasmic ferritin is seen. (E) A mature erythrocyte with Heinz bodies fixed to the red cell membrane. This cell is comparable to that shown in SEM in Fig. 2B. It is from a patient with Hb Köln. (Courtesy Dr. J. Breton-Gorius, Paris, France)

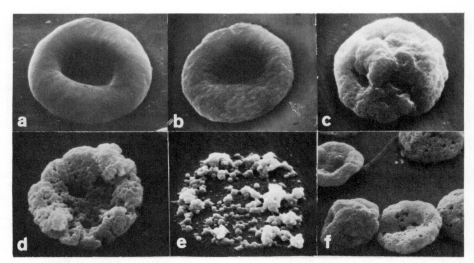

Fig. 2.—Heinz bodies, scanning electron microscopy (SEM). (A and B) Cells incubated with phenylhydrazine show gradual changes from a normal discocyte to one with multiple excrescences at the surface of the cell. (C and D) The remaining unprecipitated hemoglobin may be washed from the cell leaving a cast of hemoglobin precipitates with the general shape of the former cell. (D, E, F) After drying, the cell cast collapses to form an irregular aggregate of hemoglobin and membrane precipitates on the mounting medium.

microvascular circulatory splenic sieve and a proportion of the rigid Heinz bodies are pitted from erythrocytes by reticuloendothelial cells and removed by fragmentation;[8,27] or whole cells may undergo erythrophagocytosis or osmotic lysis. Red cell cation permeability and osmotic fragility are increased in cells with Heinz bodies from patients who are heterozygous for Hemoglobin Köln.[15] This is presumably due to mixed disulfide bonds which are formed between denatured globin (most likely involving the beta-93 cysteine) and membrane proteins. This change allows cation leak, produces ATP exhaustion and deprivation which is then followed by osmotic hemolysis. In beta thalassemia red cells show excessive potassium leak and a loss of intracellular water.[23] It has been shown that the excess alpha chains in these cells aggregate, undergo denaturation and form liaisons with the membrane in a fashion apparently similar to that described for cells with Heinz bodies.[12]

Hemoglobin C Disease

Hemoglobin C is characterized by an alteration of one polar group in an external helical region of the beta chain.[25] This theoretically increases intermolecular attraction via interaction of this relatively positively charged region with a negatively charged receptor in the alpha chain of an adjacent molecule. The molecular aggregation is somewhat favored by deoxygenation of the cell and by relatively decreased solubility of HbC.[7] Intraerythrocytic crystals may be formed in vivo and the subsequent intracellular derangement of hemoglobin to cell membrane relationship favors fragmentation and hemolysis of the cell. In nonsplenectomized patients with HbC disease, cells containing crystals are rarely found in the peripheral blood, presumably because of efficient sequestration and destruction of these rigid cells in the splenic microcirculation.[9] After splenectomy, crystal-bearing cells are more numerous and may comprise 5 per cent of the circulating cells. The decreased intracellular hemoglobin solubility is readily demonstrated by brief incubation of the cells in 3 per cent solution of sodium chloride. Under hyperosmotic conditions, the cells rapidly lose water and intracellular molecular aggregation and crystallization, which occurs at concentrations of 32-34 Gm. per cent (a concentration at which hemoglobin A remains soluble), is greatly accelerated.[7]

Intraerythrocytic hemoglobin crystals are easily identified by phase contrast or interference microscopy. Examination of cells after hypertonic incubation provides a "crystal preparation" which is useful in the diagnosis of HbC disorders (Fig. 3A,B). The crystals are birefringent and strongly absorb 414 mμ (Soret band) light and there may be two to five crystals in one cell.[21] Occasionally, one may observe smaller crystals which fuse to form a larger single crystal. Electron microscopy of freeze-fracture-replications of HbC cells indicate that 70 Å cytoplasmic particles, which are very likely hemoglobin molecules, exist in a state of aggregation (Fig. 3C,D). In comparison, the interior of a normal cell shows random dispersion of hemoglobin molecules. In both normal and HbC cells, increased molecular packing is present in the juxtamembrane regions of the erythrocyte. With progressive dehydration of HbC cells there is a further decrease in interaggregate spaces, coalescence of HbC aggregates takes place and paracrystalline alignment of the cytoplasmic particles is seen adjacent to the cell membrane. When further dehydration occurs, distinct intracellular paracrystals are seen with a periodic tetragonal array of hemoglobin molecules (Fig. 3E).[21]

The intracellular hemoglobin changes are associated with morphologic changes which are evident upon examination of the cell surface. Scanning electron microscopy (Jensen, unpublished data) of plasma-suspended cells shows irregularly shaped microcytes, microspherocytes, and biconcave cells which have deep, sometimes multiple invaginations, or less frequently, evaginations from the cell surface (Fig. 4A–F). Target cells in HbC disease are more numerous in dried smears made on glass surfaces than when plasma-suspended cells are examined. Target cells have been seen in scanning electron beam microscopy and Soret Band microscopy with a centrally located hemoglobin crystal, and others are present which have membranous folds near the center of the cell which form the central density. The latter type of target cells is seen predominantly when cells are prepared by a dried smear technique (Fig. 4B).

Hemoglobin C target cells and discocytes manifest a filterability intermediate to HbC microspherocytes and normal cells and have an increased viscosity.[7] These observations are most compatible with an increased degree of organization of the hemoglobin. The increased mechanical fragility of the HbC cells is also most likely a function of their increased internal viscosity and decreased plasticity. In addition, HbC microspherocytes have an increased

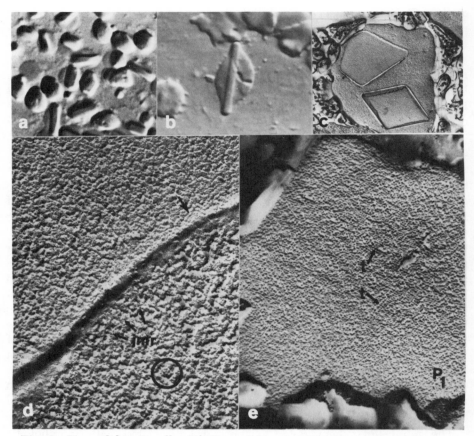

Fig. 3.—Hemoglobin C cells. (A) Hemoglobin C cells after incubation in hypertonic salt solution seen with interference microscopy. Cells are spherocytic and intra- and extracellular crystallization of hemoglobin has occurred. (B) A single intraerythrocytic rod-like crystal of hemoglobin C with cell membrane drawn about the ends of the crystal. (C) Low power electron microscopy of a freeze-etched preparation from a hemoglobin C cell demonstrating two intraerythrocytic crystals. (D) Portions of adjacent HbC cells after brief hypertonic dehydration and freeze-fracture-replication. The upper cell shows tightly packed cytoplasmic particles (hemoglobin molecules). The linear orientation of the particles in the juxtamembranous portion of the cell is seen. The lower cell shows aggregates of hemoglobin C molecules (circle) interspersed with interaggregate spaces to produce a network within the cell. Aggregates show increased packing in the juxtamembrane region (JMR). (E) Transmission electron micrograph of a portion of a HbC crystal. Periodic molecular alignment is present within this paracrystalline pattern showing a tetragonal array in certain regions (T). Individual hemoglobin particles measure 70 Å in diameter.

osmotic fragility and reach a critical hemolytic volume more rapidly than the discoid HbC cells or normal red cells.

These morphologic observations are compatible with the concept that hemoglobin aggregation and microcrystallization begin near the cell membrane and thereafter extend to form larger crystals in the less dense bulk hemoglobin of

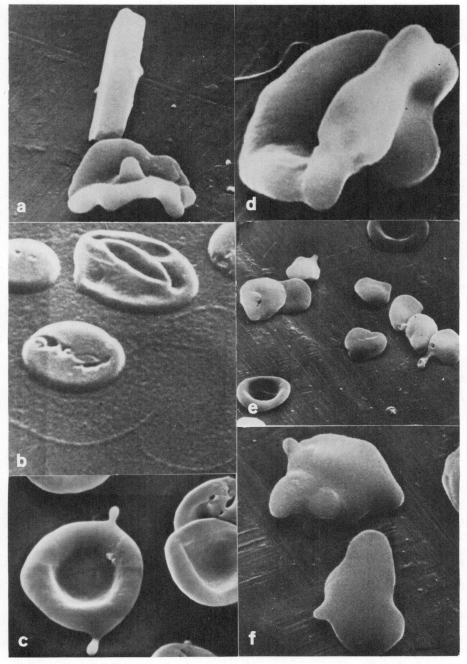

Fig. 4.—Hemoglobin C cells, scanning electron microscopy. (A) Following incubation in hypertonic medium, very large crystals may form. The target cell is of the "Mexican hat" type. The large extracellular hemoglobin crystal was probably formed within a cell. (B) This peripheral blood smear on a glass slide shows the deformities of the cell including the wrinkled central folds of the cell which forms

the cell.[21] These rigid microcrystals would, by virtue of their surface location (as with Heinz bodies), be subject to pitting or phagocytosis by reticuloendothelial cells during passage through the splenic microcirculation. It is quite possible that membrane loss occurs as a result of fragmentation with a symmetrical loss of membrane lipid and the formation of the dense microspherocytes.

Sickle Cell Disease

The major features of sickle cell disease (vascular stasis, thrombosis and hemolysis) are attributed to a single genetically determined abnormality of the primary structure of the beta chain of the hemoglobin molecule.[24] Valine is substituted for glutamic acid in the sixth amino acid from the C-amino terminus of the beta polypeptide (an external region of the A helix).[25,14] Hypothetically, hydrophobic bonding between valines in the first and sixth positions of the beta chain distorts the tertiary structure of the hemoglobin molecule.[22] Upon deoxygenation, this alteration in the external region of the first portion of the beta chain of the hemoglobin molecule allows interaction with an adjacent alpha chain receptor which permits stacking of HbS molecules into monomolecular filaments. Suggestive evidence of this interaction exists in HbS-Memphis in which an alpha chain abnormality (alpha-23 glu→gln) coexists with the conventional HbS (beta-6 glu→val) substitution.[19] The alpha chain substitution affects the receptor site and theoretically interferes with molecular stacking of the deoxygenated hemoglobin, thus interrupting the formation of monomolecular filaments of HbS. This interruption of molecular interaction greatly mollifies the otherwise expected pathologic consequences. Similarly, hybrid hemoglobin mutants containing one beta chain of HbA and one beta chain of HbS would inhibit the close ordering of molecules of deoxygenated HbS.[25]

Murayama studied cell-free solutions of HbS by the technique of whole mount electron microscopy.[22] In shadowed whole mount specimens he found rods with a diameter of 160-170 Å which appeared to be composed of six helical monomolecular filaments, each of about 70 Å in diameter. Studies of sections of sickled cells by electron microscopy showed the existence of parallel rods within the sickled cells (Fig. 5A)[2,4,10,32,35] These parallel rods form fiber bundles which are commonly oriented in the long axis of the cell but may occasionally be intertwined with other transverse fibers. These rigid fibers produce gross shape changes of the cell and are visible as linear "wrinkles" or create a sculpted pattern on the cell surface. The helical polymerization of HbS

one type of target cell. Folds of dried, fixed plasma proteins form half halos adjacent to the erythrocytes. (C) Deep narrow inversions and out pouchings of cell membrane are seen in plasma suspended cells by phase microscopy but are more evident in scanning electron microscopy. (D) The cell is grossly distorted by an intracellular crystal along the right hand edge which is covered by apparently normal membrane (SEM). (E and F) Cells from the lower layers of a centrifuged column of blood vary in shape and are smaller than normal biconcave cells. These are fragmented, angular or spherical cells.

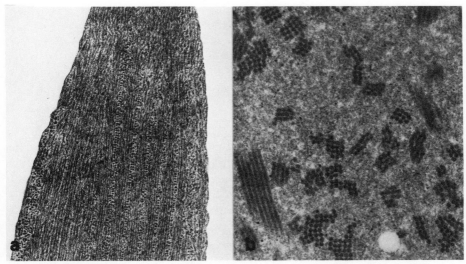

Fig. 5.—Electron microscopic sections of a sickled cell and cell-free Hb S after deoxygenation. (A) A sickle cell after deoxygenation shows the longitudinal alignment of bundles of rods of hemoglobin interspersed with transverse parallel-oriented rods. The rods measure approximately 170 Å in diameter and extend in parallel bundles to the extremity of the cell. (B) Section of deoxygenated cell-free sickle hemoglobin shows 170 Å rod-like polymers oriented into tight parallel bundles and a nearly paracrystalline arrangement in cross-section. A clear central area is suggested in cross-sections of certain rods. (Courtesy of Dr. John Bertles and Miss Joanna Döbler).

molecules is blocked by low temperature (less than $4°$ C), colchicine and vinca alkaloids, agents known to inhibit protein polymerization in formation of diverse types of microtubules.[36] Propane, under pressure, causes depolymerization of HbS helices, presumably by interference with 1-6 valo-valoyl hydrophobic bonding.[22] Sections of cell free deoxygenated HbS show bundles of rods identical to those seen in the sickled cell (Fig. 5B).[1] Electron microscopy of freeze-etched sickled cells shows rods which are 170-190 Å in diameter and are oriented in a "basketweave" pattern adjacent to the internal aspect of the cell membrane (Fig. 6C).[20] As sickling proceeds, the rods become organized into parallel fiber bundles, directed in the long axis of the body or spicule of the sickled cell (Fig. 6D). At very high resolution, by freeze-fracture-replication, the rods of 170 Å length are seen to be composed of helical strands of monomolecular filaments that are 70 Å in length (Fig. 7B). Helical rods which remain fixed to the internal surface of the membrane after freeze-fracture of the sickled cell may also be seen (Fig. 7C). This finding suggests that membrane-bound sickle hemoglobin participates early in the process of helical polymerization and is important in the biophysical, mechanical and morphologic changes of the sickled cell.

The surface of the partially sickled cell shows the cell membrane, closely adherent to a randomly oriented pattern of fibers which are about 500 Å in diameter and correspond to the intracellular parallel bundles of rods (Fig.

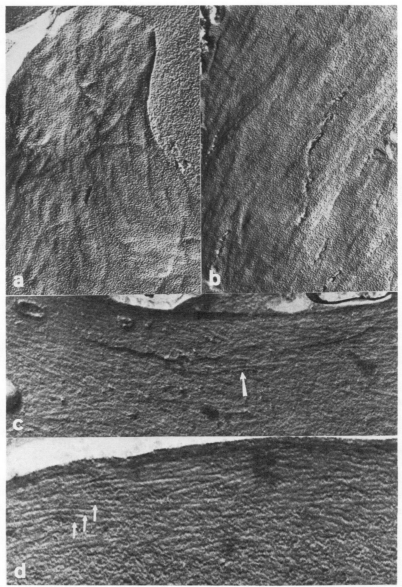

Fig. 6.—Sickle cells after freeze-fracture replication. (A) The external surface of a sickled cell after 5-minute exposure to metabisulfite. The surface fibers exhibit a random pattern. (B) The cell shows the effect of progressive sickling (after 15-minutes exposure to metabisulfite) with fibers which become oriented in a tight spiral pattern about the cell surface. (C) The interior of a cell freeze-etched after 5 minutes exposure to metabisulfite. Intracellular rods of approximately 170 Å (arrow) diameter form a basketweave pattern generally parallel to the cell membrane. Portions of vacuoles are present (left). (D) Increased sickling is associated with the formation of bundles of rods which are parallel to the cell surface, interspersed with transverse rods presenting a round or donut appearance which have a diameter of 170 Å.

Fig. 7—Freeze-etching electron micrographs of sickle cells after sickling with metabisulfite. (A) A partially sickled cell with a transverse fracture of rods of hemoglobin, 170 Å in mean diameter with five to six subunits (circles) surrounding a clear central core. (B) Magnified portion of a partially sickled cell showing a rod of hemoglobin with a helical structure. Three turns of the helix are seen (small arrows). This 170 Å rod is composed of 70 Å monomolecular filaments (upper arrow). Monomolecular filaments are in turn disposed into helical windings as shown in the model (inset). (C) A helix remaining on the internal surface of a sickled cell after freeze-fracture-replication showing branching, helical configuration and the intimate association with the internal surface of the membrane.

6A, 6B). These fibers were first identified by Bessis by means of electron microscopy of moulages of sickle cells[4] (Fig. 7). When complete sickling of the cell has occurred, surface fibers are oriented in a tight spiral about the cell (Fig. 6B). The organization of the HbS in sickled cells which causes cell surface abnormalities is well seen in the stereoscan electron microscope where deoxygenated cells show multiple bizarre forms (Fig. 8A).[17] The distortion is usually in two dimensions and only occasionally in three dimensions. The numerous long thin spicules displayed by some sickle cells suggest an abundant cell membrane. They may be compared with the "freakish poikilocytes" which commonly occur in sickle cell disease and have been termed "irreversibly sickled cells" which seem to have much less abundant cell membrane. In the oxygenated form they are bipolar, ovoid, or curvilinear and have a smooth surface (Fig. 8C). Upon deoxygenation they have approximately the same gross appearance, lack the long spicules and acquire a hard and chiseled surface (Fig. 8F). This observation and those obtained by phase microcinematography of living cells during oxygenation and deoxygenation suggest that the "irreversibly sickled cell" has insufficient membrane to allow the formation of long thin spicules. The irreversibly sickled cells are dense, have a high hemoglobin concentration and an increased viscosity. Once formed, they are rapidly removed from the circulation.[31]

There are several mechanisms whereby hemoglobin polymerization may cause cell destruction. Sickled cells have increased mechanical fragility and rigid spiculed forms such as the central sickled cell (Fig. 8A) are susceptible to damage by the turbulence and shearing forces of blood flow in large vessels as well as in the microvascular filtration process. Avulsion of the rigid cellular filaments by mechanical forces existent in the circulation could cause either intravascular hemolysis or erythrophagocytosis of the fragmented cell. A second and distinctly different mode of cell destruction may occur with the changes in shape which the cell undergoes during oxygenation or deoxygenation. In the sickled cell the normal hemoglobin-to-membrane relationship is greatly distorted; one portion of the cell contains a thick mass of organized hemoglobin and other portions are relatively depleted of hemoglobin. This allows the inner surfaces of the cell membrane to come into close apposition while the cell is in the sickled state. During unsickling, fusion of the apposed membranous surfaces takes place, followed by compartmentalization of the cell and, thereafter, by fragmentation (Fig. 8B,D,E). The loss of cell fragments is apparently associated with a symmetrical loss of membrane and the eventual creation of a cell with an increased hemoglobin concentration, increased density, increased osmotic and mechanical fragility and unusual distorted shapes (poikilocyte).[17] Irreversibly sickled cells have a decreased surface area as compared with normal cells or young discocytes from patients with sickle cell anemia. Irreversibly sickled cells have been described by Milner and Serjeant[31] as short-lived cells and by Bertles[3] as dense, viscous cells with greatly reduced plasticity or deformability.

Analyses of the deoxygenated solutions of mixtures of HbS, HbA, HbC and HbF show that HbA and HbC participate in the formation of rods seen in sickled cells but that fetal hemoglobin does not and is excluded from the

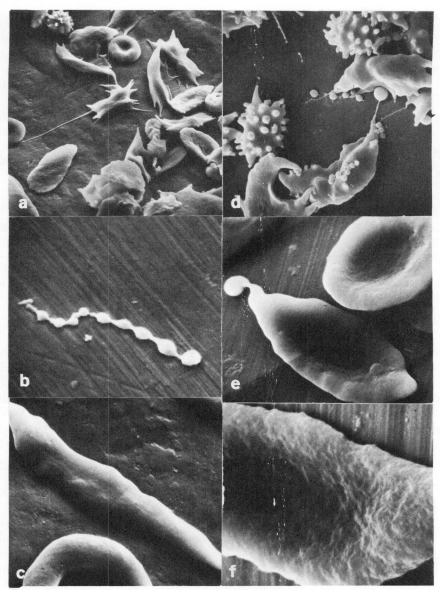

Fig. 8.—Sickle cells. (A) Deoxygenated sickled cells have bizarre shapes with long filaments, short spicules or sculpted and chiseled surfaces. There are thin, veil-like, membranous portions in some cells where hemoglobin is virtually absent. In other parts of the cells there is a dense accretion of hemoglobin. (B, C, D). During the unsickling process, hemoglobin in spicules liquifies, the bulk hemoglobin flow occurs toward the main cell and frequently spherules of membrane-enclosed hemoglobin are entrapped. They may remain attached to the cell by short necks or thin myelin strands. (E and F) Repetitive loss of membrane and hemoglobin with multiple sickle and unsickle experiences, may lead to membrane loss and the formation of the "irreversibly sickled" poikilocyte which, in the oxygenated state, is a smooth-surfaced, elongated, cigar-shaped cell and in deoxygenated state shows a sculpted surface which is characteristic of the polymerized intracellular hemoglobin.

Fig. 9.—Bone marrow cells (SEM). (A) Deoxygenation shows a distortion of the surface of nucleated red cells and reticulocytes which contain sufficient hemoglobin to allow polymerization. The spherical cells with irregular surface membranous protrusions are leukocytes. (B) The rounded acorn-like sphere is a deoxygenated extruded red cell nucleus surrounded by residual sickle hemoglobin which is enclosed in membrane. The surface has the appearance of sickling without the formation of filaments.

molecular interaction.[1] Irreversibly sickled cells have small amounts of HbF as compared with other cells in the blood of patients with sickle cell anemia.[3] This might be a result of the sickle-unsickle process if HbF is excluded from the intracellular polymerization. It may be more available as the soluble hemoglobin contained within the fragments which are shed from the cell during the unsickling process. By this mechanism, the ratio of HbS to HbF in a cell would be expected to increase as fragments are lost and the cell is converted to an irreversibly sickled cell. Deoxygenated HbS undergoes polymerization in bone marrow reticulocytes and hemoglobinized normoblasts (Fig. 9A,B). The sickling of residual perinuclear hemoglobin in a membrane enclosed extruded nucleus is shown in Fig. 9B.

The fragmentation induced by the unsickling process in vitro is not unique to the human sickle cell.[18] The red cells of the deer *(Cervidae)* which sickle with oxygenation (rather than deoxygenation) display the fragmentation phenomenon during the deoxygenation unsickling process.[17] Sickling in deer cells begins with a characteristic sickle deformation, progressing to the "matchstick" appearance in which the membrane of the red cell is stretched around a long central hemoglobin rod. The rod can be seen on freeze-fracture microscopy to be crystal-like on transverse section and is in close contact with the red cell membrane (Fig. 11).

The finding of fragmentation when shape change is induced by polymerization or depolymerization of the protein in two entirely different molecular species of hemoglobin (HbS and deer hemoglobin) allows the suggestion that the phenomenon is dependent upon the shape change and membrane properties, rather than a specific species of hemoglobin. These observations

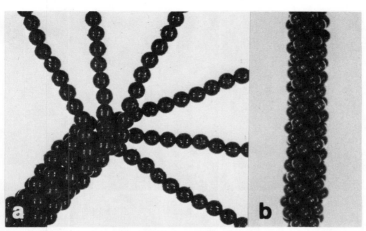

Fig 10.—Models of the hypothetical molecular mechanism of sickling (after Murayama.) (A) Deoxygenation of Hb S permits stacking of molecules into linear filaments. Monomolecular filaments twist into a loose alpha helix. The helices twist into fixed strands which tighten into rod-like polymers with complete sickling. (B) Morphologic evidence for this process is presented in Fig. 7B.

strongly suggest that the inner surfaces of the cell membrane are capable of interaction and that fusion occurs between juxtaposed portions of the inner surface of the cell membrane. This property would allow egress of effete nonfunctional organelles (such as the nucleus) from the cell without loss of bulk hemoglobin. The same phenomenon, the annealing of internal membrane surfaces, may be necessary to the phagocytic bisection of an erythrocyte to yield hemoglobin-replete spherocytes and to the formation of poikilocytes in such diverse disorders as intravascular coagulation, iron deficiency anemia and valve prosthesis hemolytic anemia. In all of these examples, interior cell surface membrane apposition would be provided by extracellular forces, whereas in the hemoglobinopathies and deer cells, the cell deformation which leads to cell membrane interaction is produced by an intracellular reversible change in the physical state of the abnormal hemoglobin.

Summary

Three examples of red cell disorders have been presented in which abnormal hemoglobins interact with the cell membrane and cause premature destruction of the cell.

Heinz body formation, which occurs in the red cells of some patients with unstable hemoglobins, is initiated with the oxidation of normal hemoglobin to methemoglobin and hemichromes, and is followed by the aggregation of irreversibly oxidized and denatured hemoglobin. Morphologically, the progression of Heinz body formation can be followed in the cell from the appearance of aggregates of small numbers of molecules to macroaggregates which form a liaison with the internal surface of the cell membrane. This causes mem-

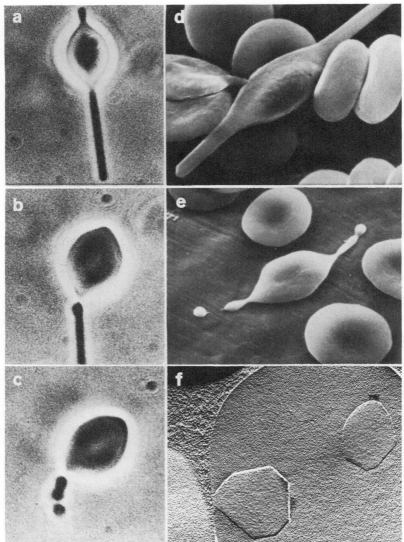

Fig. 11.—Shape changes in deer cells (phase contrast and SEM). (A, B and C) Oxygenation of deer cells *(Cervidae)* at pH 6.7, causes formation of the sickled "matchstick" forms. The plasma suspended single cell seen undergoes unsickling to form a discocyte with fragmentation of the cell and formation of microspherules of hemoglobin. These frames were taken from a 16 mm. cine film. (D, E and F) Surface morphology (SEM) shows membrane-enclosed globules of hemoglobin formed during unsickling of the cell. The crystalline nature of the hyperoxygenated hemoglobin in the cell is demonstrated in the freeze-etch photograph (F).

brane rigidity and the abnormal cells are either disrupted or partially phago-cytized in their passage through the microcirculation.

In HbC disease, the second example, there is increased intermolecular attraction and aggregation with subsequent protein crystallization. The greater degree of organization of hemoglobin molecules within the cell produces increased internal rigidity and decreased plasticity of the HbC cell. The decrease in cell plasticity and relative excess of red cell membrane allows the pitting or fragmentation of paracrystals from portions of the cell. Fragmentation and/or partial phagocytosis leads to production of microspherocytes and greatly deformed poikilocytes which have shortened circulatory life spans.

In the third disorder, sickle cell anemia, the abnormal deoxygenated hemoglobin molecules become stacked into monomolecular filaments which twist into helical rods which are associated to form parallel fiber bundles. The fiber bundles distort and elongate the sickled cells, rendering them rigid and incapable of normal passage through the microcirculation. The rigidity and increased mechanical fragility result in cell disruption or phagocytosis. Cells also undergo cycles of sickling and unsickling and exhibit fragmentation and membrane loss which may result in formation of the "irreversibly sickled" poikilocyte. High resolution morphologic studies suggest an intimate relationship between polymerized rods of sickled hemoglobin and the internal surface of the red cell membrane. As hemoglobin unsickles within the long central rod of the cell, fragmentation occurs. Thus, hemoglobin instability and denaturation or the polymerization of inherited abnormal hemoglobins, because of their close, possibly even structural association with the red cell membrane, lead to hemolysis by a variety of interrelated mechanisms. Common to all are partial phagocytosis, fragmentation and premature destruction of the cell.

ACKNOWLEDGMENT

We thank Mrs. Karen Olson for technological assistance and Miss Joyce Schofner for her help with the preparation of the figures.

REFERENCES

1. Bertles, J. F.: Hemoglobin interaction: modification of solid phase composition in the sickling phenomenon. Science (in press) 1970.

2. Bertles, J. F., and Döbler, J.: Reversible and irreversible sickling: a distinction by electron microscopy. Blood 33:884, 1969.

3. Bertles, J. F., and Milner, P. A.: Irreversibly sickled erythrocytes: a consequence of the heterogeneous distribution of hemoglobin types in sickle cell anemia. J. Clin. Invest. 47: 1731, 1968.

4. Bessis, M., Nomarski, G., Thiéry, J. P., and Breton-Gorius, J.: Études sur la falciformation des globules rouges au microscope polarisant et au microscope électronique. II. L'intérieur du globule. Comparison avec les cristaux intra-globulaires. Rev. Hématol. 13:249, 1958.

5. Carrell, R. W., and Lehmann, H.: The unstable haemoglobin haemolytic anemias. Seminars Hemat. 6:116, 1969.

6. Carson, P. E., and Frischer, H.: Glucose-6-phosphate dehydrogenase deficiency and related disorders of the pentose-phosphate pathway. Amer. J. Med. 41:744, 1966.

7. Charache, S., Conley, C. L., Waugh, D. F., Ugoretz, R. J., Spurrell, R. J., and Gayle, E.: Pathogenesis of hemolytic anemia in hemozygous hemoglobin C disease. J. Clin. Invest. 46:1795, 1967.

8. Crosby, W. H.: Normal functions of the spleen relative to red blood cells: review. Blood 14:399, 1959.

9. Diggs, C. W., Kraus, A. P., Morrison, D. B., and Rudnicki, R. P.: Intraerythrocytic crystals in a white patient with hemoglobin C in the absence of other types of hemoglobin. Blood 9:1172, 1954.

10. Döbler, J., and Bertles, J. F.: The physical state of hemoglobin in sickle cell

anemia erythrocytes in vivo. J. Exp. Med. 127:711, 1968.

11. Fessas, P., Loukopoulos, D., and Kaltsoya, A.: Peptide analysis of the inclusions of erythroid cells in β-thalassemia. Biochim. Biophys. Acta 124:430, 1966.

12. Fessas, P., Loukopoulos, D., and Therell, B.: Absorption spectra of inclusion bodies in β-thalassemia. Blood 25:105, 1965.

13. Heinz, R.: Morphologische Veränderungen der rotten Blut Körperchen durchen Gifte. Virchow Arch. [Path. Anat.] 122:112, 1890.

14. Hunt, J. A., and Ingram, V. M.: Allelomorphism and the chemical difference of the human haemoglobins A, S, and C. Nature (London) 181:1062, 1958.

15. Jacob, H. S., Brain, M. C., Dacie, J. V., Carrell, R. W., and Lehmann, H.: Abnormal haem binding and globin SH group blockade in unstable haemoglobins. Nature (London) 218:1214, 1968.

16. Jandl, J. H., Engle, L. K., and Allen, D. W.: Oxidative hemolysis and precipitation of hemoglobin. I. Heinz body anemias as acceleration of red cell aging. J. Clin. Invest. 39:1818, 1960.

17. Jensen, W. N.: Fragmentation and the "freakish poikilocyte." Amer. J. Med. Sci. 257:355, 1969.

18. Kitchen, H., Putnam, F., and Taylor, W. J.: Hemoglobin polymorphism in white tailed deer: Subunit basis. Blood 29:867, 1967.

19. Kraus, L. M., Miyaji, T., Iuchi, I., and Kraus, A. P.: Characterization of $\alpha^{23}gluNH_2$ in hemoglobin Memphis: hemoglobin Memphis/S a new variant of molecular disease. Biochemistry (Wash.) 5:3701, 1966.

20. Lessin, L. S.: Polymerisation hélecoidole des molecules d'hémoglobine dans les erythrocytes falciformes. Étude par cryodécapage. C. R. Acad. Sci. [D] (Paris) 266:1806, 1968.

21. Lessin, L. S., Jensen, W. N., and Ponder, E.: Molecular mechanism of hemolytic anemia in homozygous hemoglobin C disease. J. Exp. Med. 130:443, 1969.

22. Murayama, M.: Molecular mechanism of sickling. Science 153:145, 1966.

23. Nathan, D. G., Stossel, T. B., Gunn, R. B., Zarkowsky, H. S., and Laforet, M. T.:

Influence of hemoglobin precipitation on erythrocyte metabolism in alpha and beta thalassemia. J. Clin. Invest. 48:33, 1969.

24. Pauling, L., Itano, H. A., Singer, S. J., and Wells, I. C.: Sickle cell anemia, a molecular disease. Science 110:543, 1949.

25. Perutz, M. F., and Lehmann, H.: Molecular pathology of human haemoglobin. Nature (London) 219:902, 1968.

26. Rifkind, R. A., and Danon, D.: Heinzbody anemia, ultrastructural study. I. Heinz-body formation. Blood 25:885, 1965.

27. Rifkind, R. A.: Heinz-body anemia, ultrastructural study. II. Red cell sequestration and destruction. Blood 26:433, 1965.

28. Rigas, D. A., and Koler, R. D.: Decreased erythrocyte survival in hemoglobin H disease as a result of the abnormal properties of hemoglobin H: the benefit of splenectomy. Blood 18:1, 1961.

29. Rosenberg, S. A., and Guidotti, G.: The proteins of the erythrocyte membrane: Structure and arrangement in the membrane. Red Cell Membrane: Structure and Function. In Jamieson, G. A., and Greenwalt, T. J. (Eds.): Red Cell Membrane Structure and Function. Philadelphia, J. B. Lippincott, 1969, p. 93.

30. Selwyn, J. F.: Heinz bodies in red cells after splenectomy and after phenacetin administration. Brit. J. Haemat. 1:173, 1955.

31. Serjeant, J. R., Serjeant, B. E., and Milner, P. F.: The irreversibly sickled cell: a determinant of haemolysis in sickle cell anaemia. Brit. J. Haemat. 17:527, 1969.

32. Stetson, C. A., Jr.: The state of hemoglobin in sickled erythrocytes. J. Exp. Med. 123:341, 1966.

33. Weed, R. I., and Reed, C. F.: Membrane alterations leading to red cell destruction. Amer. J. Med. 41:681, 1966.

34. Weed, R. I., and Weiss, L.: The relationship of red cell fragmentation occurring within the spleen to cell destruction. Trans. Ass. Amer. Physicians 79:426, 1966.

35. White, J. G.: The fine structure of sickled hemoglobin in situ. Blood 31:561, 1968.

36. White, J. G., and Krivit, W.: Induction of sickling phenomenon in erythrocytes exposed to glutaraldehyde. J. Cell Biol. 35:141A, 1967.

Index